Invertebrate Welfare

Invertebrate Welfare

Guest Editors

William Winlow
Jennifer Mather
Anna Di Cosmo

MDPI

Basel • Beijing • Wuhan • Barcelona • Belgrade • Novi Sad • Cluj • Manchester

Guest Editors
William Winlow
Department of Biology
University of Naples Federico II
Naples
Italy

Jennifer Mather
Department of Psychology
University of Lethbridge
Lethbridge
Canada

Anna Di Cosmo
Department of Biology
University of Naples Federico II
Naples
Italy

Editorial Office
MDPI AG
Grosspeteranlage 5
4052 Basel, Switzerland

This is a reprint of the Special Issue, published open access by the journal *Animals* (ISSN 2076-2615), freely accessible at: https://www.mdpi.com/journal/animals/special_issues/PAC3X4I96N.

For citation purposes, cite each article independently as indicated on the article page online and as indicated below:

Lastname, A.A.; Lastname, B.B. Article Title. *Journal Name* **Year**, *Volume Number*, Page Range.

ISBN 978-3-7258-7316-6 (Hbk)
ISBN 978-3-7258-7317-3 (PDF)
https://doi.org/10.3390/books978-3-7258-7317-3

Cover image courtesy of Michaella P Andrade

Contents

About the Editors

William Winlow

William Winlow is an Emeritus Professor of Neuroscience, associated with the University of Naples Federico II in Italy. He has experience of working on crustacea (PhD, St Andrews) and molluscs (NYU with Eric Kandel), and then with Paul Benjamin (University of Sussex), and has also worked on mice for a short period of time at Glasgow University with Peter Usherwood. He taught Physiology and Neuroscience at Leeds University for 23 years. His major interests include neuroethology, the function of neural networks, and anesthetic mechanisms. He is currently working on neurocomputational mechanisms with Andrew Johnson. He has substantial editorial experience.

Jennifer Mather

Jennifer Mather is an Emeritus Professor in the Department of Psychology at the University of Lethbridge in Canada. She has studied the behavior of cephalopods, primarily octopuses, for over 50 years. Her papers have included ethological descriptions of behavior in the field, as well as play, personalities, and problem solving in the lab. Her study of the potential consciousness of cephalopods led to her concern for their welfare and, from there, broadened to advocacy for other invertebrate groups as well.

Anna Di Cosmo

Anna Di Cosmo, Full Professor of Zoology at the University of Naples Federico II, is a leading expert in cephalopod neurobiology and head of OctopusCore, Italy and Europe's only authorized cephalopod research facility. Her research has revealed the neurophysiological bases of reproductive behavior, adult neurogenesis in the octopus brain, and the sensory capabilities of arms and suckers, which detect light via opsins even in the optic lobes. She has inspired studies showing how octopuses perceive chemical cues through touch, revealing a sophisticated sense of "smell by touch." She also developed a unique anesthesia protocol for octopuses, combining $MgCl_2$ as a pre-anesthetic muscle relaxant with isofluorane as a true anesthetic. Di Cosmo played a key role in including cephalopods in EU Directive 2010/63, the first legal recognition of invertebrate sentience, and actively promotes invertebrate welfare through research, outreach, and interdisciplinary advocacy.

animals

MDPI

Editorial

Ethics and Invertebrates: The Problem Is Us

Jennifer A. Mather

Department of Psychology, University of Lethbridge, Lethbridge, AB T1K 3M4, Canada; mather@uleth.ca

Simple Summary: Why do we need a collection of papers about the welfare of invertebrates, and what do we need to learn from it? The most important reason is that they make up most of the animals on the planet, so animal welfare without invertebrates simply is not the welfare of animals. Being vertebrates and mammals, we have concentrated on knowing about animals like us, and so there is a huge lack of information about invertebrates. We cannot care about them if we do not know they exist. More than that, we do not particularly like them. They are often small, seen as ugly cockroaches, parasitic ticks, biting mosquitoes, but think instead of dazzling butterflies, flexible octopuses, attractive fireflies. Most of all, we do not think about their welfare because we see them as just machines, no complex behavior and intelligence, not deserving of care. Recent science research is changing that, and we are beginning to see that they are often very smart and sensitive and deserve our consideration. Why this is true and how we can extend welfare to non-vertebrates is the subject of this set of papers.

Abstract: In the last few decades, science has begun to make great strides at understanding how varied, fascinating, and intelligent invertebrate animals are. Because they are poorly known, the invertebrates that make up about 98% of the animals on the planet have been overlooked. Because they are seen as both simple and unattractive, children and their teachers, as well as the general public, do not think they need care. Because until recently we did not know they can be both intelligent and sensitive—bees can learn from each other, butterflies can navigate huge distances, octopuses are smart, and lobsters can feel pain—we have to give them the consideration they deserve. This collection of papers should help us to see how the lives of invertebrates are tightly linked to ours, how they live, and what they need in terms of our consideration and care.

1. Introduction

Why does the situation of invertebrate welfare start with us? Because the biggest problems with invertebrate welfare are centered around humans, not the animals themselves. We have come to dominate the planet during the present epoch, now deservedly called the Anthropocene [1]. The damage we have done to earth and the animals on it is likely caused by the early Western philosophical approach to these animals, starting with the *scala naturae*, the belief in ascending improvement leading to humans as the peak of evolution [2]. This viewpoint was picked up by the Christian church with its belief that 'man shall have dominion', and due to Descartes' assertion that animals were mere things to be manipulated, distinct from humans who had souls.

Of course, this is not the only ethical approach to animals. Buddhists believe in the brotherhood of all animals and the transformation of living things after death, along with the continuity of energy [3]. Additionally, the plains Indians believed in the one-ness of nature and our brotherhood with animals [4]. But, the Christian view became the dominant one in science. This anthropocentrism caused three problems for invertebrates: a lack of knowledge, negative attitudes, and a misunderstanding of their cognitive abilities, all leading to problems for welfare consideration. These played out through the interacting trio of (1) *scientific knowledge*, (2) *beliefs of the general public*, and (3) *political/commercial interests*.

2. We Do Not Know Much about Invertebrate Animals

In terms of the scientific lack of knowledge about invertebrates, Schleffers et al. [5] discuss what the term 'missing biodiversity' means. Mammals comprise 0.5% of the animal species on the planet, and yet we have named and identified 97% of them, but probably only half of the molluscs and one third of the crustaceans. This is particularly true of animals in unstudied ecosystems such as oceans, which make up 70% of the habitable space on the planet. This lack of information has led to the decade-long *Census of Marine Life,* a major attempt by science to find some of these animals. It has been a huge advance, but we are still catching up. This gap in information is partly the result of neglect by scientists, as an examination of recent papers in the journal *Animal Behaviour* [6] showed that 1/3 of papers were on these mammals and only 21% were on insects, who represent 170 times as many species (1). Further, papers in leading animal welfare journals have focused nearly completely on care for mammals. The result is that legislation protecting the welfare of invertebrates in research exists in only a few countries in the world, particularly in the UK, European Union, and almost only for cephalopods [6], with the United States as a notable exception.

The lack of scientific knowledge is reflected in that of the general public. Kellert [7] found that the American public knew little about invertebrates, answering only 1/3 of simple questions about them correctly. Others [8] looked at French students in Environmental and Human Sciences and found that mammals were much better known and that insects were not even seen as part of the environment. Golick [9] checked university students in the US for mistakes about insects and found them frequent. Since Kellert [7] had found that those with more education in general had a better knowledge of animals, this lack of understanding by university students makes the [9] study even more worrying.

Media feed the vertebrate centrism in our information about animals, and the public responds. Scientists have to know about animals, but they also have to feed into and inform the rest of us. The Paris Zoological Park had an 'adoption' program for the public to specifically target individual animals with donations. Of the 29 possible species, 20 were mammals and only one, the tarantula, was an invertebrate [10]. A similar adoption program at the Vancouver Aquarium featured three animals, two mammals and a shark, even though there are 130 marine mammals and 200,000 identified fish species. No charismatic invertebrates were featured. Albert et al. [11] performed a wide survey of the media using both an internet survey and data from aquariums and Disney and Pixar films, and found that 20 species were emphasized in these media, 17 of which were large mammals. Tourists visiting sub-Saharan Africa have been keyed to look for the 'big five' large mammals and their conservation is thought to support the preservation of the ecosystems of the region [12], though this excuse is only partly true. Only some of the ecosystems on the continent are preserved from human expansion and exploitation. Even when other authors [13] have explicitly said that they wanted to study anthropomorphism in the evaluation of animals, they managed to only include three invertebrates in their total of 30 species. Their view did not widen very much.

3. We Do Not Like or Value the Invertebrates That We Do Know about

Given this lack of information about invertebrates, perhaps it is not surprising that people have negative evaluations of them, especially of insects. They are seen as ugly, crunchy/squishy, even menacing. In fact, Schlegel et al. [14] found that despite this unpleasant view, children who could recognize them had less negative attitudes to invertebrates, so it is partly due to ignorance. This was the same with Yugoslavian university students, who had mixed views of insects, despite the fact that some of them had just taken a course on bee biology [15]. Kellert [7] pioneered the study of attitudes to animals by interviewing 29 Americans, and he found the invertebrates (examples of which were cockroach, mosquito, and wasp) occupied three of the four 'least liked' positions, accompanied by the rat as third-least, with affective views of fear and disgust. Given the relationship of these species to humans, this is perhaps not surprising; he could have instead chosen butterflies, dragonflies, and shrimp, for instance. A later author [16] looked at attitudes to non-human animals in terms of positions sorted on different dimensions and found four clusters in how we catalogue them: food species, popular species, intelligent ones, and dangerous ones. This could have been followed up to evaluate the attitudes to members of the different groups but it was not. We could feature 'popular' invertebrates such as butterflies, but we mostly do not. The negative attitudes start with children, as Swiss children showed disdain for land invertebrates [17], though they were more positive to butterflies and dragonflies [14]. In fact, butterflies were admired by residents and tourists in northern India [18] and noticed by visitors to urban parks in East and Southeast Asia [19]. Breuer et al. [17] specifically studied them as 'flagship' species for invertebrates in Beverin Nature Park in Switzerland, aiming to help turn attitudes around. The more general disdain can start early, as [20] found this preference in Italian kindergarten children. However, this is not surprising as both [21,22] found them showing fear and disgust toward invertebrates, and especially toward spiders.

It is not surprising that children have negative attitudes to invertebrates when you look at the values of their teachers. An evaluation of students in teacher education in Norway and Italy showed that there were negative attitudes toward invertebrates, especially in Italy and much more in females [23]. This gender gap is worrying, as early education teachers are predominantly female, and prejudices learned early are hard to counter. Still, the cultural difference shows that these negative attitudes are learned and not innate. This is particularly true as 'outdoor education' is a feature of Norwegian life and children meet and interact much more with invertebrates than they would if they were indoors playing on the computer. The dislike and ignorance are not just European, as there is a low level of knowledge and understanding of invertebrates in Indonesia too [24]. A short experience is not always enough to change sweeping attitudes, as exposure to Madagascar cockroaches changed pre-teacher attitudes to that insect but did not generalize to other species [22].

However, it is relatively easy to turn this ignorance and dislike around somewhat in children. German school children's visits to a 'green classroom' outside the school had a long-term effect on them noticing that invertebrates existed [25]. This was shown when they drew pictures of what lived there, as they now especially noticed the soil and its inhabitants. Japanese gardeners had biophilia (liking) to insects and not the more common biophobia (aversion) found in their society [26], again having experienced the ecosystem 'close up'. Many American elementary schools have gardens in the schoolyard for children to work in, a break from, and extension of, the classroom. This began with the goal of increasing their likelihood of eating a healthier diet, especially including vegetables [27], assuming that the involvement and good taste would win them over. Tending to the soil seems to have a mental health outcome as well, with the change in focus and the productivity helping with stress. However, given the Japanese study, it may increase children's knowledge of, and comfort with, invertebrates as a side effect.

4. We Do Not Know How Smart and Sensitive They Are

Because of our anthropocentric view of animals, we have long underestimated the intelligence of any invertebrate group. The dominant belief in the evolution of cognitive ability is the social intelligence hypothesis [28,29]. Focusing particularly on primates, researchers have argued that the pressure to understand other individuals in tight-knit social groups has led to the flexibility of actions and knowledge of consequences. They particularly talked about the 'theory of mind', the necessity to understand the motivation of others. The mammalian, and especially primate, combination of big brains, long life, parental care, and these complex social groups have fostered learning and its use. Still, the evaluation of hyenas argued for a modified version [30], and others saw a combined ecological-social one [31], but the paradigm was fairly fixed, dominated the study of animal cognition, and generally eliminated invertebrates from any considerations of intelligence, sentience, and thus welfare.

As a result of this view, several biases have interfered with a fair assessment of invertebrate abilities and their need for welfare consideration [32]. One is the belief that different structures and systems cannot be convergent for mental operations. Mammals have a cortex and consciousness, so no cortex means no consciousness. A second is that size (big brains) is a dominant influence and small brains cannot carry out the complex operations needed for intelligence. A third is the mammal-centered use of testing animals for cognitive ability within situations in the sensory modality and adaptation for the structures of vertebrates. The 'mark test' of Gallup [33], demanding visual dominance and non-living skin structure such as fur, is the most obvious. Finally, there is the excessive use of Morgan's canon, the belief that any simple explanation of the source of behaviour must be the correct one if it is held up against a more complex one. The answer is 'usually but not always'.

As more evidence accumulated about the intelligence of cephalopod molluscs in particular [34,35], it became apparent that these invertebrates were very intelligent. This was brought to the more general attention of scientists by the Cambridge Declaration of Consciousness [36]. A gathering of leading neuroscientists and behaviorists evaluated animals with the aim to find possible neural substrates for consciousness. Among other groups, octopuses were suggested as candidates for this ability, and this judgment was widely circulated. A philosopher's semi-popular book about cephalopods' cognitive ability [37] reached many in the scientific community, and a more subjective and popular one about the author's experiences with them [38] in the ocean and in captivity was read by the general public. Craig Foster's documentary *My Octopus Teacher* [39], besides winning an Oscar, reached all across the globe, and gradually resulted in a huge expansion of public information and sympathy for them, thus slowly changing people's attitudes. Popular opinion about how the intellectual ability of cephalopods could have evolved is still varied, leading a group of cosmologists to ignore the biological evidence and suggest that octopuses were aliens who had actually descended from space as cryopreserved eggs [40].

The scientific presentation of the possibility of insect consciousness [41] has fared less well. The authors take a neurobiological approach to proof of consciousness and suggest that the seat of vertebrate consciousness may be not the cortex but the midbrain, allowing them to evaluate insect consciousness 'on a more level playing field'. As they expected when they produced these ideas, there were what they called "lively areas of disagreement". As Michalevich and Powell [32] suggested, opponents to the consideration of insects as intellectual advanced the size problem, disbelieved the analogy, and made the call for retreat to simplicity in explanations. Additionally, there was much discussion of the relative roles of the cortex and midbrain in consciousness. Research continues to 'chip away' at the exclusivity of complex behaviour as a mammalian domain, and the disdain for insects' cognitive competence. See [42] for insect navigation, [43] for social cognition, and [44] for a suggestion of a direct comparison between insects and vertebrates; however, changing a paradigm is slow work. In terms of the public presentation of this scientific information, the recently published short book Mind of a Bee [45] is aimed at a wider audience. It is a start, but further developments will be needed to convince the public.

This collection of papers will be an introduction to why and how we should 'do' invertebrate welfare. It is not directed specifically at the general public, but at those seeking education on this wider view of animal welfare. There is a complex web of scientists, public opinions, and institutions. This includes 'farmers' who raise domestic species, governments who regulate what we can and must do to invertebrates, and NGOs who advocate for animals. But here is where and with whom it starts.

Acknowledgments: I would like to thank my co-editor, Bill Winlow, for getting me into this.

Conflicts of Interest: The author declares no conflict of interest.

References

1. Laurance, W.E. The Anthropocene. *Curr. Biol.* **2019**, *9*, R953–R954. [CrossRef]
2. Mather, J.A. Philosophical background of attitudes towards and treatment of invertebrates. *ILAR J.* **2011**, *52*, 205–212. [CrossRef]
3. Barash, D.P. *Buddhist Biology*; Oxford University Press: Oxford, UK, 2010.
4. Booth, A.L.; Jacobs, A.L. Ties that bind: Native American beliefs as a foundation for environmental consciousness. *Environ. Ethics* **1990**, *12*, 27–43. [CrossRef]
5. Schleffers, B.R.; Joppa, L.N.; Pinn, S.L.; Laurance, W.F. What we know and don't know about earth's missing biodiversity. *Trends Ecol. Evol.* **2012**, *27*, 501–510. [CrossRef]
6. Mather, J.A. What is in an octopus' mind? *Anim. Sentience* **2019**, *26*, 1–29.
7. Kellert, S.R. Values and perceptions of invertebrates. *Conserv. Biol.* **1993**, *7*, 845–855. [CrossRef]
8. Leandro, C.; Jay-Robert, P. Perceptions and representations of animal diversity: Where did the insects go? *Biol. Conserv.* **2019**, *237*, 400–408. [CrossRef]
9. Gollick, D.A.; Heng-Moss, T.M.; Steckleberg, A.L.; Brooks, D.W.; Higley, L.G.; Fowler, D. Using web-based key character and classification instruction for teaching undergraduate students insect identification. *J. Sci. Educ. Technol.* **2013**, *29*, 509–521. [CrossRef]
10. Colleony, A.; Clayton, S.; Couvet, D.; Saint Jaime, M.; Prevot, A.-C. Human preferences for species conservation: Animal charisma trumps conservation status. *Biol. Conserv.* **2005**, *8*, 263–269.
11. Albert, F.; Luque, G.M.; Courchamp, F. What are "Charismatic species" for conservation biologists? *Biosci. Master Rev.* **2016**, *10*, 1–8.
12. Sergio, F.; Carp, T.; Clucas, B.; Hunter, J.; Ketchum, J.; Hiraldo, F. Top predators as conservation tools: Ecological rationale, assumptions, and efficacy. *Ann. Rev. Ecol. Evol. Syst.* **2008**, *39*, 1–19. [CrossRef]
13. Eddy, T.J.; Gallup, G.G.; Povinelli, D.J. Attribution of cognitive states to animals: Anthropomorphism in comparative perspective. *J. Soc. Issues* **1993**, *49*, 87–101. [CrossRef]
14. Schlegel, J.; Breuer, G.; Rumpf, R. Local insects as flagship species to promote nature conservation? A survey among primary school children on their attitudes to invertebrates. *Anthrozoos* **2015**, *28*, 229–245. [CrossRef]
15. Stanisavljevic, J.D.; Stanisavljevic, L.Z. Attitudes of university students of biology towards bees and their protection. *J. Biosci. Biotech.* **2017**, *6*, 215–219.
16. Driscoll, J.W. Attitudes toward animals: Species ratings. *Soc. Anim.* **1995**, *3*, 139–150. [CrossRef]
17. Breuer, G.; Schlegel, J.; Kauf, J.; Rupf, R. The importance of being colorful and able to fly: Interpretations and implications of children's statements on selected insects and other invertebrates. *Int. J. Sci. Educ.* **2015**, *37*, 2664–2687. [CrossRef]
18. Barua, M.; Gurdak, D.J.; Ahmed, R.A.; Tamuly, J. Selecting flagships for invertebrate conservation. *Biodivers. Conserv.* **2012**, *21*, 1457–1476. [CrossRef]
19. Lim, V.C.; Sing, K.W.; Chong, K.Y.; Jaturas, N.; Dong, H.; Lee, P.S.; Tao, N.T.; Le, D.T.; Bonebrake, T.C.; Tsang, T.P.; et al. Familiarity with, perceptions of, and attitudes towards butterflies of urban park users in mega-cities across East and Southeast Asia. *Soc. Open Sci.* **2022**, *9*, 220161. [CrossRef]
20. Borgi, M.; Cirulli, F. Attitudes towards animal among kindergarten children: Species preferences. *Anthrozoos* **2015**, *28*, 45–59. [CrossRef]
21. Nageotte, N.; Buck, C. "I gotta touch that?" Attitudes and self-efficacy of pre-service teachers regarding scary or disgusting science. *J. Outdoor Environ. Educ.* **2020**, *23*, 173–190. [CrossRef]
22. Wagler, A.; Wagler, R. Beliefs about future curriculum: How avoidance emotions affect curriculum choice in science. *Soc. Anim.* **2016**, *24*, 596–617. [CrossRef]
23. Melis, C.; Falcicchio, G.; Wold, R.A.; Billing, A.M. Species identification skills in teacher education students: The role of attitude, context and experience. *Int. J. Sci. Educ.* **2021**, *43*, 1709–1725. [CrossRef]
24. Madang, K.; Tibrani, M.M.; Santoso, L.M. Analysis of pre-service biology teachers' metacognitive skills on invertebrate zoology. *Math. Sci. Educ. Int. Sem.* **2019**, *1731*, 012006. [CrossRef]
25. Drissner, J.R.; Haase, H.-M.; Wittig, S.; Hille, K. Short-term environmental education: Long-term effectiveness? *J. Biol. Educ.* **2013**, *48*, 9–15. [CrossRef]
26. Vanderstocke, A. For the love of insects: Gardening grows positive emotions (biophilia) towards invertebrates. *J. Insect Conserv.* **2022**, *26*, 751–762. [CrossRef]
27. Graham, H.; Beall, D.L.; Lussier, M.; McLaughlin, P.; Zidenberg-Cherr, S. Use of school gardens in academic institutions. *J. Nutr. Educ. Behav.* **2005**, *37*, 145–151. [CrossRef]
28. McNally, L. Cooperation and the evolution of intelligence. *Proc. R. Soc. B* **2012**, *279*, 3027–3034. [CrossRef]
29. Johnson-Ulrich, L. The social intelligence hypothesis. In *Encyclopedia of Evolutionary Psychological Science*; Springer: Berlin/Heidelberg, Germany, 2017.
30. Holekamp, K. Questioning the social intelligence hypothesis. *Trends Cogn. Sci.* **2007**, *11*, 65–69. [CrossRef]
31. Henke-van den Marlsberg, J.; Kappeler, P.; Fichtel, C. Linking ecology and cognition: Does ecological specialization predict cognitive test performance? *Behav. Ecol. Sociobiol.* **2020**, *74*, 154. [CrossRef]
32. Michalevich, I.; Powell, R. Minds without spines: Evolutionarily inclusive animal ethics. *Anim. Sentience* **2020**, *29*. [CrossRef]
33. Gallup, G.G. Self-awareness and the evolution of social intelligence. *Behav. Proc.* **1998**, *42*, 239–247. [CrossRef] [PubMed]

34. Mather, J.A.; Dickel, L. Cephalopod complex cognition. *Curr. Opin. Behav. Sci.* **2017**, *16*, 131–137. [CrossRef]
35. Amodio, P.; Boeckle, M.; Schnell, A.K.; Ostajic, L.; Fiorito, G.; Clayton, N.S. Grow smart and die young: Why did cephalopods evolve intelligence? *Trends Ecol. Evol.* **2019**, *34*, 45–56. [CrossRef] [PubMed]
36. Low, P.; Edelman, D.; Koch, C. Consciousness in human and non-human animals. In Proceedings of the Francis Crick Memorial Conference, Cambridge, UK, 7 July 2012.
37. Godfrey-Smith, P. *Other Minds: The Octopus, the Sea, and the Deep Origins of Consciousness*; William Collins: New York, NY, USA, 2016.
38. Montgomery, S. *The Soul of an Octopus*; Atria Books: New York, NY, USA, 2020.
39. Foster, C. My Octopus Teacher; Documentary. Available online: https://www.imdb.com/title/tt12888462/ (accessed on 1 September 2023).
40. Steele, E.J.; Al-Mufti, S.; Augustyn, K.A.; Chandrajith, R.; Coghlan, J.P.; Coulson, S.G.; Ghosh, S.; Gillman, M.; Gorczynski, R.M.; Klyce, B.; et al. Cause of Cambrian explosion—Terrestrial or cosmic? *Prog. Biophys. Mol. Biol.* **2018**, *136*, 3–23. [CrossRef]
41. Klein, C.; Barron, A.B. Insect consciousness: Commitments, conflicts and consequences. *Anim. Sentience* **2016**, *9*, 21. [CrossRef]
42. Collett, M.; Chittka, L.; Collett, T.S. Spatial memory in insect navigation. *Curr. Biol.* **2013**, *23*, R789–R800. [CrossRef]
43. Chittka, L.; Rossi, N. Social cognition in insects. *Trends Cogn. Sci.* **2022**, *22*, P578–P592. [CrossRef] [PubMed]
44. Simons, N.; Tibbets, E. Insects as models for studying the evolution of animal cognition. *Curr. Opin. Insect Sci.* **2019**, *34*, 117–122. [CrossRef]
45. Chittka, L. *The Mind of a Bee*; Princeton University Press: Princeton, NJ, USA, 2023.

animals

MDPI

Editorial

Postscript to Invertebrate Welfare: "We Have Met the Enemy and He Is Us"

William Winlow [1,2,*], Jennifer Mather [3,*] and Anna Di Cosmo [1,*]

1 Department of Biology, University of Naples Federico II, 80126 Naples, Italy
2 Institute of Ageing and Chronic Diseases, University of Liverpool, Liverpool L69 3BX, UK
3 Department of Psychology, University of Lethbridge, Lethbridge, AB T1K 3M4, Canada
* Correspondence: bill.winlow@gmail.com (W.W.); mather@uleth.ca (J.M.); anna.dicosmo@unina.it (A.D.C.)

Through this collection of papers, we have considered in depth the effects that humans have on invertebrate welfare in a variety of contexts. For instance, we have discussed how these effects manifest in breeding or display facilities and in testing and research laboratories, as well as in the wild. The articles comprising this collection, though few, span species of different phyla and in conditions, yet they represent only a sample of contexts relevant to this research topic. To facilitate this discussion, we have grouped these scenarios into representative tables. For those who know little about invertebrates, References [1,2] provide ample background for a sufficient understanding.

1. How Humans View Invertebrates (See Table 1)

The field of animal welfare is an inclusive one, containing ideas from philosophy, ethics, law, agriculture and fisheries, tourism—including zoos and aquaria—and even veterinary care. World Animal Protection defines welfare as follows: "The physical and psychological well-being of an animal. It is good or high if the individual is fit, healthy, free to express natural behavior, free from suffering and in a state of wellbeing." It is likely that the first clear definition of the term was that of the Brambell Committee in the UK, which investigated the welfare of agricultural animals, leading to the recognition of the often cited five freedoms [3]: freedom from (a) hunger and thirst, (b) discomfort, (c) pain, injury, and distress, (d) and fear, and (e) freedom to express normal behaviour. However, some have reported that these factors are too focused on negative freedom; Dawkins [4] suggested that domestic animals have a high degree of plasticity, and that we should evaluate not their 'natural behaviour' but what the animals themselves value as experiences. Mellor [5], supported by Miller and Chinnadurai [6], proposed that we focus instead on five provisions: these include good nutrition, a satisfactory environment, good health, appropriate behaviour, and positive mental experiences. In either case, what these two definitions share is the necessity to evaluate whether animals are doing well.

By virtue of living in the Anthropocene era [7], everything on and in the earth is strongly affected by human attitudes and actions. Tracking human attitudes is paramount, as the public view of ethics changes when new information is revealed, and human views necessarily influence scientific practice [8].

Table 1. Studies on how humans view invertebrates.

• Ethics and Invertebrates: The Problem Is Us, by **Jennifer A. Mather**
• The Long Road from Religious and Ethical Traditions to Welfare of Invertebrates, by **Jennifer A. Mather**

2. Attitudes to Captive Invertebrates (See Table 2)

Ethical concerns, public opinion, legal regulations, and material values all work together to 'keep us in the dark' about invertebrate welfare [9]. And yet the biggest problem we face with regard to this field of study is the lack of available information about invertebrates and their need for welfare consideration. What are we doing (or not doing) to them? In what ways are we influencing them? What do we need to do for them?

Table 2. Research on attitudes toward captive invertebrates.

• Invisible Invertebrates: The Welfare of Invertebrates in Public Aquaria, by **Kerry Perkins**
• Assessing the Welfare of Captive Group-Housed Cockroaches, *Gromphadorhina oblongonota*, by **Danielle Free** and **Sarah Wolfensohn**
• Ethical Considerations for Echinoderms: New Initiatives in Welfare, by **Augusto César Crespi-Abril** and **Tamara Rubilar**

It is of course most important to consider the welfare of invertebrates when they are our captives and when we have direct control of them. A portion of this collection features situations involving captive invertebrates, including those on display and those used in science. Similar contexts that we were not able to discuss in this collection include other ecotourism practices, such as insectariums and butterfly farms [10]. A more extensive emerging concern is the agricultural use of insects, with already one trillion insects being used as human food [11].

3. Care of Captive Invertebrates (See Table 3)

The 3R principles of welfare—reuse, reduce, and replace—[12] are sometimes interpreted to mean "replace the important and sentient mammals in testing with unimportant invertebrates". However, modern research advances in animal cognition have made it clear that complex information processing is not just the territory of vertebrates. Cephalopod molluscs and decapod crustaceans in particular have been shown to be much more intelligent and sensitive than we previously assumed [13].

Table 3. Studies on pain and anaesthesia.

• Behavioural Indicators of Pain and Suffering in Arthropods and Might Pain Bite Back? by **Robert W Elwood**
• The Use of Isoflurane and Adjunctive Magnesium Chloride Provides Fast, Effective Anaesthetization of Octopus vulgaris, by **Anna Di Cosmo**, **Valeria Maselli**, **Emanuela Cirillo**, **Mariangela Norcia**, **Heethaka K. S. de Zoysa**, **Gianluca Polese** and **William Winlow**
• Effects of acetic acid and morphine in shore crabs, *Carcinus maenas*: implications for the possibility of pain in decapods, by **Stuart Barr and Robert W. Elwood**

It is surely the responsibility of humans, and particularly of scientists, to take the welfare of invertebrates seriously and to treat them well in captivity and in the wild. Because of this, we ought to focus on both pain and awareness, and how we might measure and prevent them. Other papers have zeroed in on wider solutions; for example, Crook [14] focuses on the future responsibilities of young researchers and veterinary groups [15].

4. Human Influences on Wild Invertebrates (See Table 4)

Although our initial and continuing focus has been on the welfare of animals under direct human control, this has to expand. One direction of expansion has gone beyond the direct human influence and control of animals in captivity to the indirect influence we have on 'wild' animals across ecosystems globally. In the Anthropocene era [7], human influence is pervasive throughout all regions on Earth. This includes the inadvertent and planned modification of the physical environment and the unplanned effects that come with this, such as tourism, pollution, and climate change. These factors have significant real-life consequences, regardless of whether they were planned, and they matter to animals outside of our direct control.

Table 4. Papers on invertebrates in the wild.

• Effects of Artificial Light at Night on Fitness-Related Traits of Sea Urchin (*Heliocidaris crassispina*), by **Xiuwen Xu, Zexianghua Wang, Xiuqi Jin, Keying Ding, Jingwen Yang** and **Tianming Wang**
• Assessing Negative Welfare Measures for Wild Invertebrates: The Case for Octopuses, by **Michaella P. Andrade, Charles Morphy, D. Santos, Mizziara M. M. De Paiva, Sylvia L. S. Medeiros, C. E. O'Brien, Françoise D. Lima, Janaina F. Machado** and **Tatiana S. Leite**
• It Is Not Only Data—Freshwater Invertebrates Misused in Biological Monitoring, by **Paweł Koperski**
• Increasing Risks to the Health of the Invertebrates—Balancing between Harm and Benefit, by **Tatiana V. Kuznetsova, Valentina A. Kudryavtseva** and **Larisa L. Kapranova**

Because the scope of this research topic is so wide, we have been able to feature research from diverse backgrounds. These include accidental pollution, inadvertent mistreatment, and the careless use of chemicals and capture devices. It was also necessary to expand the scope of this collection to include the individual influences of light pollution [16] and chemical toxicity [17]. In addition, we have added references regarding their effects on invertebrates in whole land [18] and marine [19] ecosystems. Furthermore, we must not ignore the effects of temperature increase in this period of climate change, which produces hypoxic and acidic marine and freshwater environments, which are deleterious to aquatic life [20,21].

In our activity, we affect the whole planet [7], and invertebrates make up 98% of the animals on and in it. Of course this matters. In conclusion, "**We have met the enemy and he is us**" [22].

Conflicts of Interest: The authors declare no conflicts of interest.

References

1. Prather, C.M.; Pelini, S.L.; Laws, A.; Rivest, E.; Woltz, M.; Bloch, C.P.; Del Toro, I.; Ho, C.; Kominoski, J.; Newbold, T.A.S.; et al. Invertebrate ecosystem services and climate change. *Biol. Rev.* **2013**, *88*, 327–348. [CrossRef] [PubMed]
2. Lewbart, G.A.; Zachariah, T.T. Aquatic and terrestrial invertebrate welfare. *Animals* **2023**, *13*, 3375. [CrossRef] [PubMed]
3. Brambell Committee. *Report of the Technical Committee to Inquire into the Welfare of Animal Kept under Intensive Livestock Husbandry Systems;* Hansard Parliament: London, UK, 1965; Volume Voume 718.
4. Dawkins, M.S. Farm animal welfare: Beyond "natural" behavior. *Science* **2023**, *379*, 326–328 [CrossRef] [PubMed]
5. Mellor, D.J. Moving beyond the "five freedoms" by updating the "five provisions" and introducing aligned "Animal Welfare Aims". *Animals* **2016**, *6*, 59. [CrossRef] [PubMed]
6. Miller, L.J.; Chinnadurai, S.K. Beyond the five freedoms: Animal welfare at modern zoological facilities. *Animals* **2023**, *13*, 11318. [CrossRef] [PubMed]
7. Laurance, W.E. The Anthropocene. *Curr. Biol.* **2019**, *9*, R953–R954. [CrossRef] [PubMed]
8. Drinkwater, E.; Robinson, E.J.H.; Hart, A.G. Keeping invertebrate research ethical in a landscape of shifting public opinion. *Methods Ecol. Evol.* **2019**, *10*, 1265–1273. [CrossRef]
9. Cardoso, P.; Erwin, T.L.; Borges, P.A.; New, T.R. The seven impediments in invertebrate conservation and how to overcome them. *Biol. Conserv.* **2011**, *144*, 2647–2655. [CrossRef]
10. Lemelin, R.H.; Boileau, E.Y.S.; Russell, C. Entomotourism: The allure of the arthropod. *Soc. Anim.* **2019**, *27*, 733–750. [CrossRef]
11. Barrett, M.; Adcock, S. Animal welfare science: An integral piece of sustainable insect agriculture. *J. Insects Food Feed.* **2024**, *10*, 517–531. [CrossRef]
12. Russell, W.M.S.; Burch, R.L. *The Principles of Humane Experimental Technique;* Methuen: London, UK, 1959.
13. Birch, J.; Burn, C.; Schnell, A.; Browning, H.; Crump, A. Review of the evidence of sentience in cephalopod molluscs and decapod crustaceans. In *Report of the London School of Economics;* LSE: London, UK, 2021.
14. Crook, R.J. The welfare of invertebrate animals in research: Can science's next generation improve their lot? *PostDoc J.* **2013**, *1*, 9–18. [CrossRef]
15. Wahltinez, S.J.; Stacy, N.I.; Hadfield, C.A.; Harms, C.A.; Lewbart, G.A.; Newton, A.L.; Nunamaker, E.A. Perspective: Opportunities for advancing aquatic invertebrate welfare. *Front. Vet. Sci.* **2022**, *9*, 973376. [CrossRef]
16. Rodrigo-Comino, J.; Seeling, S.; Seeger, M.K.; Ries, J.B. Light pollution: A review of the scientific literature. *Anthr. Rev.* **2023**, *10*, 367–392. [CrossRef]
17. El-Din, M.I.S. A review of the effects of pharmaceuticals on marine invertebrates. *Biomed. J. Sci. Tech. Res.* **2023**, *48*, 40196–40198.

18. Saunders, M.E. Ecosystems services in agriculture: Understanding the multifunctional role of invertebrates. *Agric. For. Entomol.* **2018**, *20*, 298–300. [CrossRef]
19. Chen, E.Y.-S. Often overlooked: Understanding and meeting the current challenges of marine invertebrate conservation. *Front. Mar. Sci.* **2021**, *8*, 690704. [CrossRef]
20. Gobler, C.J.; Bauman, H. Hypoxia and acidification in ocean ecosystems: Coupled dynamics and effects on marine life. *Biol. Lett.* **2016**, *12*, 20150976. [CrossRef] [PubMed]
21. Janes, T.A.; Syed, N.I. Evolutionary sophistication of aerial respiratory behaviour in the freshwater mollusc *Lymnaea stagnalis*. In *Neuroecology and Neuroethology in Molluscs—The Interface between Behaviour and Environment*; Di Cosmo, A., Winlow, W., Eds.; Nova: Hauppauge, NY, USA, 2014; pp. 177–2010.
22. Kelly, W. *Pogo Cartoon, Earth Day Poster*; Smithsonian: Washington, DC, USA, 1970.

animals

MDPI

Commentary

The Long Road from Religious and Ethical Traditions to Welfare of Invertebrates

Jennifer A. Mather

Department of Psychology, University of Lethbridge, Lethbridge, AB T1K 3M4, Canada; mather@uleth.ca

Simple Summary: The welfare of invertebrates is a result of ethical human behaviour, presently stemming from religious traditions, but how do we get from religious traditions to ethical behaviour to the welfare of invertebrates? The Judaeo–Christian religions are firmly anthropocentric, though they urge utilitarian care for domestic animals. Similarly, the Buddhist tradition of rebirth, possibly as an animal, leads to a belief in ahisma or doing no harm. The theory does not necessarily lead to practice in either case, and Indigenous societies are not usually hunter-gatherers nowadays. Religious practices are often close to good ecology and lead to better consideration. None of these traditions extends much to invertebrates, as only those invertebrates thought 'worthy' of consideration are ones that might be sentient. Philosophy is often not based on biological reality, and science demands the simplest possible explanation and 'objective factual truth.' The result is a long haul to find and apply the best ethical principles to the welfare of these invertebrates, which are the majority of animals on the planet.

Abstract: Ethical behaviour tends to lead to the welfare consideration of animals, but much less so for invertebrates. Indigenous tradition often valued all animals as having an important role in life on the planet, a practical application of modern ecology. The Judaeo–Christian–Islamic tradition postulated 'man' as having dominion over all of Earth, resulting in anthropocentrism and careless practices. In contrast, the Buddhist/Hindu belief in rebirth leads to ahisma, or doing no harm. In the face of capitalist systems, practice does not necessarily follow these beliefs, especially in the 'shepherding' of domestic animals. Only Jainist beliefs value the lives of all invertebrates. Philosophers are often divorced from the physiological reality of the animals they muse about, and science's traditions of objectivity and the simplest possible explanation of behaviour led to ignorance of invertebrates' abilities. Ninety-seven percent of animals on the planet are invertebrates. We have a long way to go to provide moral standing and welfare consideration, which is consistent with the new information about the sentience of some of these animals.

Keywords: invertebrates; welfare; ethics; religious values; moral standing; sentience

1. Introduction

Ethics must be behind ideas about invertebrate welfare, but it is a long, rocky road to get there. The start has to be some definitions—realising, alas, that they are English and Western. The Cambridge Dictionary defines ethics as "The study of what is morally right or wrong". The word comes from the Greek ethos, meaning habit, encompassing duties and also consequences. Morality, according to the same source, is "a set of personal or social standards for good or bad behaviour". McKay and Whitehouse [1] look for the psychological basis of such a set of standards. They see it as biologically derived and evolved, based on a 'cognitive architecture' or way of thinking. At the same time, a set of moral values is a culturally produced group of religious and moral representations. They believe that all sets of values look for and judge 'fairness' or unbiased evaluation, yet at the same time, all of them use what they call 'agency detection', or seeking to find out why things happen and finding different explanations. People tend to evaluate actions not only on the basis of fairness but also in affective terms [2]. Every group begins to see themselves as a whole, forming an ingroup identity and fostering kinship detection. You will pass on your genes to related individuals, so it is important to know and value the related us vs. strangers. This results in the formation of a culture-specific religious/moral identity and a set of rituals that have only social meaning. Think of Jewish prayer shawls, the Navaho dawn song, Christian Sunday, and Buddhist prayer wheels, all getting meaning from and giving meaning to those who perform or use them. Through this group identity, morality becomes all bound up with religion. Singer [3] was explicit about this when he said, "Religious traditions have historically been the principal vehicle by which the status of non-human animals was evaluated". In action, religious practices are usually the foundation of morality, especially within larger cultural groups [4].

2. Religious Traditions about the Welfare of Animals

Since they are foundational, it is useful to explore what different religious traditions have defined and codified as morality with respect to animals. There are several, and their values are different [5]. The Abrahamic traditions, including Islam, are most familiar to us Westerners, so familiar that we are sometimes in danger of seeing them as 'the' values. Singer [3] and Szucs et al. [6] point out that this pattern of tradition stresses anthropocentrism, that man is both special and separate from animals, and in the Bible, the book of Genesis explicitly says, "Man shall have dominion over. . .. every living thing that moveth upon the Earth". This parallels Aristotle's idea of the scala naturae, the concept that evolution moved from rocks to plants to 'lower' animals and on to humans as the pinnacle of evolution, just below angels. This explicit separation was later fostered in philosophy by Descartes, who believed that animals were mere machines and that we could do anything we liked to, in contrast to humans, who had souls. The Qur'an says, "Then we made you heirs in the land after them" [7]. Such a separation of us from nature persists to this day in Eurocentric societies and has been used to justify the exploitation of land and animals that Western society is based on. Even in science, there is a long history of trying to find a characteristic that sets us apart, disproven one by one. First, there was 'man the social species', then 'man the language user' (but see bee dances), then 'man the tool user', and when that one ran out, 'man the tool-maker', until we saw chimpanzees and even New Caledonian crows shaping plant material to make useful tools for getting food and/or water. Such a sought-after separation seems necessary to justify exploitation.

The precepts of Buddhism and Hinduism, the second major set of religious traditions, are quite different. It is fundamental to the Buddhist tradition to do no harm, a principle called ahimsa [8]. This obviously extends to not harming animals, especially as it is believed that when we die, we are reborn in another body, possibly that of a non-human animal. All our actions are submitted to a judging process and are seen to produce and store up either good or bad karma. With accumulated bad karma, you are reborn as a 'lower' form of life, likely a non-human animal, with good karma returning as a better-off human, especially as a man. Thus, Buddhists and Hindus have several reasons to treat animals well, especially not to kill them, leading in general to a vegetarian diet. It is not clear that they realised the extent of the animal kingdom, though, or whether they should not harm invertebrates. The Jain sect did so [6], however, and extended their consideration so far as to sweep the path in front of them to avoid hurting insects and avoid eating root vegetables because it would disturb and hurt the small animal fauna of the soil.

The belief systems of 'aboriginal' or first inhabitants of many areas of the world, often delineated by differences from settler invaders, are varied but have common themes. Perhaps the best explained in terms of attitudes towards animals are those of North American First Nations [9]. They see power and influence in every part of the ecosystem, even including the rocks. According to the Lakotah Sioux, every animal is important in itself and in its place in the ecosystem, though they did not have that word. Perhaps because they originated as gatherer-hunters, they saw animals as special. They feel a kinship with animals and, in some cases, believe that prey species such as fish were once 'people' but transformed into food animals to feed humans (say the West Coast Indians, who explicitly called them to come upriver). Australian Aboriginal groups see the country as a mosaic of clans, territories, and kin groups, identifying strongly with their traditional land and in kinship with particular totem animals, which they never hunt [10]. As gatherer-hunters, aboriginal groups often chase and kill food animals, needing the source of protein, but respect them for 'giving themselves so the people can eat'. According to the Inuit, if the animals are not respected, they will not make themselves available to be killed. Standing Bear, a Lakotah, speaks of the joy and wonder at the elements and the seasons (something muted by our air conditioners and central heating) and of Earth as mother and Sky as father. Although they were not consciously ecologists, consideration of the welfare of other animal species was almost automatic to these peoples, who lived close to the land.

3. From Religion to Practice

To be fair to Western religious traditions, most tell their followers to be kind to the domestic animals under their control, although this is not extended to invertebrates. Care and the killing of domestic animals, is often carefully regulated, and Frayne [11] points out that humans are seen as not so much owners as caretakers of animals. The Jewish kosher dietary tradition is a kind of moral discipline of one's basic drives, and remember that ingroup formation leads to rituals such as not mixing meat and milk that give 'meaning' to what one does. Sharia law likewise dictates halal practices during the keeping of domestic animals as well as their better-known slaughter practices. Deliberately taking the life of an animal is surrounded by rituals in Jewish, Muslim, and First Nations practices to give the animal a swift end and to acknowledge its life as a gift to us; see Grandin and Regenstein's [12] detailed explanation to 'meat scientists.' This utilitarian approach to welfare may naturally accompany domestication, as what is good for the animals produces good meat and other products. In a backward kind of parallel to the drive to be ethical and care for domestic animals, research suggests that people who are abusive to animals as children may abuse other humans in adulthood [13]. These two areas also call attention to the fact that ethical behaviour may be seen to stem from a justice (what is right) and/or a caring (what is kind) approach.

Nevertheless, what the religious traditions dictate may not be what happens in daily life, even though a sample of individuals from a wide variety of countries all said they cared about animals and thought regulation of their welfare was needed [14]. Rollin [15] argues that initially, the husbandry of domestic animals, though utilitarian in focus, resulted in welfare for the animals because care and attention were good for their prosperity. 'Shepherding' as protection from predators, an ample supply of appropriate or at least fairly good food, and even rudimentary medical care were all very good for the cows, sheep, or chickens, even though they might end up killed for food. Care measures good for survival may not align with the anima's larger welfare because they do not fit with the natural history of the species concerned. Industrialisation and the profit motive have changed that [16]. Around 20% of broiler chickens, despite living very short lives, have gait problems that may indicate they are in pain. Remedies such as giving them less crowding, better feed, and longer periods in the dark would result in slower growth and, thus, less profit [17]. The knowledge that raising egg-laying hens in wire cages was poor welfare resulted in a popular concern for them, leading to free movement, including time outdoors [18]. Provision of good welfare is true only for some hens, with 'marketisation' of welfare leading to 'free range eggs' (and when did you ever see an egg ranging anywhere?) selling for a higher price [19]. Still, it was this lack of regulation that led animal rights activists to campaign, eventually leading to the Five Freedoms, though there are gaps in how welfare legislation is eventually carried out; see [20] for Australia as an example. Conversely, in countries with Buddhist roots, animal welfare is often lacking, though it 'catches up with us', so to speak. Really bad welfare practices seen in areas such as the Wuhan 'wet market' can be seen as bad karma in the Gita [21], leading to the retribution of COVID-19. Chinese people individually believe that animals ought to be protected, but

Chinese laws are lacking [22]. Rahman [7] points out the multiple situations in which fish are suffering in aquaculture in the Middle East. Rollin [15] lists the suffering of cattle in India, where they are theoretically sacred. Even in the subsistence farms of Tibet, some lip service is paid to Buddhist practices [23], but for the most part, animals are raised and slaughtered for food. It is a damning indictment.

4. Can Science Partner with Philosophy to Promote Animal Welfare?

Such a partnership sounds logical, yet it does not always succeed. Fraser [24] critiques on the basis of philosophical attitudes, and Webb et al. [25] look more at the scientists. Philosophers may look at general principles rather than details of what specific animal groups can do; the evaluation of invertebrates might be a good example of this. Godfrey-Smith [26] called on the reafference principle to look at the feedback of action paired with copies of movement commands, yet such feedback is not in the cephalopod brain, as control is quite local in the arms of the octopuses he was discussing. Philosophers may lump together unrelated groups, so Mikhalovich and Powell's [27] emphasis on the welfare of 'invertebrates', while an important step forward, could include animals of over 30 phyla, though they acknowledge the problem. The new British regulations spearheaded by the report on invertebrate welfare [28], while ground-breaking, include decapod crustaceans from the Arthropods and Cephalopods from the Molluscs, very different animals who will require very different care and protection [29]. Theorists may not recognise real progress; many welfare-oriented philosophers rail against research practices ('vivisection') that are long out of date and end up looking as if they oppose all research.

Still, there are problems when scientists address ethical issues [30]. Scientists are dedicated to objectivity, to facts rather than opinions, and reject ethics and 'feelings'. Jane Goodall was initially castigated for giving her chimpanzee subjects names because she saw them as individuals. Science has long followed Morgan's Canon, the belief that the simplest possible explanation of a behavioural phenomenon must be the right one, squashing the possible complexity of motivations even though they might be the correct ones ("It must be reflexes"). Donald Griffin famously 'came out' in the 1970s to advocate that we should recognise animals' subjective experiences [30], but it took a long time for this attitude to permeate studies of animal behaviour. Science is supposed to be objective, so practitioners do not see their own biases; the 'mirror test', ref. [31], which was supposed to evaluate animals' cognitive ability, depends heavily on visual self-identity and grooming, eliminating animals from sentience that do not visually self-evaluate. We are wary of 'animal welfare activists', remembering PETA's destructive anti-research campaign, even freeing captive animals totally unable to live in the wild. Ultimately, regulations that are good for the animals may make research more difficult to carry out; in the US, the IACUCs (Institutional Animal Care and Use Committees) are not loved but tolerated. For researchers in invertebrates, the three Rs (Refine, Reuse, and Replace) of welfare in animal research are regarded warily because Replace is sometimes interpreted as "replace those nice, sensitive mammal subjects with invertebrates to whom you can do anything you like". In the United States, which has no ethical regulation of invertebrate welfare, invasive and potentially cruel neuroscience research can be carried out. A just-published Japanese investigation of octopus sleep chopped off half of all the arms so that the animals could not pull out the electrodes that were recording brain activity [32] and then commented that it was studying 'normal' sleep cycles.

Philosophers and scientists agree, however, on the difficulty of understanding animal consciousness or subjective experience. The Cambridge Dictionary is not much help in defining it, saying it is "the state of being awake, aware of what is around you, and able to think". The difficulty is that subjective experiences are within oneself and, except for being reported by language, cannot be known outside the individual. Regardless of one's approach, whether and how to consider animals' welfare has been based on whether the species in question is sentient [27,29] and thus aware of its present condition and choices for the future. Carruthers [33] evaluates possible explanations for whether animals have Qualia, defined as "instances of subjective conscious experience", and concludes that no theory explains this very well, though the Global Workspace Theory of Baars [34] comes closest for him. Dawkins [35] takes a more practical approach in our search for phenomenal consciousness in animals as a basis for welfare activity. She evaluates different possible actions to search for consciousness but also notes that it might be better to solve more obvious animal welfare problems. A utilitarian approach may get the animal more benefit, and animals who 'want' some environmental condition do not necessarily need consciousness. As we learn more about non-vertebrates, we understand more about their possible sentience and will then need to abandon the present anthropocentrism.

Still, this is an intriguing and relevant problem, and scientists have approached it with many different techniques, measuring physiological responses and behavioural actions. Pain is a particular problem for welfare as it is important to any animal and yet is perceived to include sensory, cognitive, and difficult-to-know affective aspects [36]. Sneddon et al. [37] lay out a series of possible investigations that might evaluate pain in animals, including sensory abilities, connections to the brain, and areas within the brain that process such information. This evaluation is difficult, as physiological responses such as hormone changes may not be the same across diverse animals not related to humans, and aversive behaviour is also likely to be species-typical. Behaviour can at least give us a window into what the animal is 'thinking'. One promising approach to measuring preference is to give the animal choices, such as Crook's [38] supply of analgesia to wounded octopuses in a specific location that they previously avoided, although this proves cognition is not affected. Another is to set up a situation where animals are 'pessimistic' and see whether that biases their evaluations of ambiguous choices, as when Bateson et al. [39] 'shook' bees before giving them a difficult discrimination with an opt-out choice. A third approach is measuring how hard an animal will work to get the condition that it seems to prefer. Still, Dawkins [35] points out that seemingly difficult choices may not even need a conscious decision, so we must be careful not to make sweeping assumptions about awareness.

5. What Does All of This Ethical Background Have to Do with Invertebrates

As religious and moral philosophies seldom even recognise that invertebrates existed as animals, they give little background for studies of the application of invertebrate welfare, and even a detailed assessment of what kind of harm we do to animals [40] is aimed only at vertebrates. It is ironic that the careful understanding that many of the aboriginal peoples had of the land and its non-human inhabitants led to their greater consideration of them. Yet their traditions, which often resemble those recommended by modern ecologists, apparently did not also influence us. Perhaps that is where the modern 'objectivity' of science, which countered an impression of aboriginal beliefs as subjective and based on 'stories', held researchers back.

Philosophers in the Western tradition are still often anthropocentric in their evaluations. They see the link between moral status and ethics; if you harm a being that has moral status, that is morally wrong [41] and unethical. However, this moral status is often human-centred, as Christian tradition says that humans are the only ones 'made in the image of God' and all have Formal Moral Status (FMS), regardless of the individual's cognitive capacity. Animals are assessed for moral status by how they measure up to humans to gain possible status, and sentience and phenomenal consciousness, rather than intelligence, are the criteria [42]. In the face of this, Birch [28] acknowledges the difficulty of knowing and feels that if you cannot specify the ethical priority for a particular animal species, you should invoke the precautionary principle and treat it as well as you can anyway.

Given the difficulty of defining this private subjective experience whose possession should lead to moral standing, a more biological or neural approach should give a firmer basis for judgement. Approaching the problem from a brain perspective, Roth [43] believed in the convergent evolution of intelligence (though not sentience), comparing the mammalian frontal cortex, the insect mushroom bodies, the pallium of birds and fish, and the vertical lobe of cephalopods as parallel higher-order control areas. The neuroscience/cognitive background has led to a greater understanding of these characteristics. The Cambridge Declaration on Consciousness, written by a gathering of neuroscientists and animal behaviourists, was the first to include cephalopods as potentially sentient. Birch et al. [28] used this neural basis for sentience when explicitly advising the British government to set regulations for the welfare of not only cephalopods but also decapod crustaceans. Further on, Pennartz et al. [44] suggest a set of behaviours—qualia richness, situatedness, intentionality, integration, and a balance of dynamic changes and stability—as criteria for consciousness. Assessments of different situations that cephalopod species have reacted to by Mather and Andrade [45] show a similar behavioural array for 'mind' in this group. This analysis of invertebrates' capacity for sentience can be extended to one view: the belief that having a mental life should give an animal moral standing and that cognitive ability, brain size, and capacity to suffer pain should dictate whether we extend moral consideration to invertebrate groups [27].

Some people have separated good welfare from this struggle to gain moral standing. Broom [29] says that "all animal life should be respected". Dawkins [35] outlines the kinds of things we would like to know about animal consciousness but concludes that making their lives better should come before philosophy. She is also in favour of using animals' own choices of situations to determine what is best for them as the most valid assessment (and see [38]). With advances in research on some invertebrate groups, we are gaining a better understanding of their behavioural and neural abilities. This has led to better welfare consideration for those being included in the 'sentience club' [46], but how do we know what abilities and how much of them should lead to inclusion in 'the club'? The anthropocentric view of sentience has not led to greater clarity of this designation, and instead, an ethic of care based on each particular animal's needs is more appropriate. For this inclusive ethic, we can go all the way back to Jeremy Bentham [47], who began the outline of the utilitarian principle as 'the greatest happiness for the greatest number', and who wrote in 1789 his oft-quoted sentence: "The question is not "Can they reason?" nor "Can they talk", but" Can they suffer?".

6. Conclusions

Where does this obviously cursory overview of the journey from religion through ethics to invertebrate welfare take us? First, as pointed out earlier in this paper and in the Editorial, we lack knowledge about invertebrates. This problem is only slowly being solved, but we cannot advance welfare without this understanding. We do not even know what invertebrates live on the planet, as seen earlier in this paper, or how we should behave ethically towards this array of different groups [27]. We need to understand the physiology and behaviour of different species to choose methods to increase welfare and decrease pain. We need to understand the sentience of invertebrates [28] if we are to use it as a benchmark for deciding which animals need special care and consideration. Insects are a particular problem as they represent 60% of all animal species, and we face a huge increase in their use as food without knowledge of their care requirements to ensure their welfare [46]. Here, especially, we see the second problem: misinformation and devaluation due to the anthropocentrism fueled by the Abrahamic religious tradition and human greed. In the face of these two problems, the papers in this collection help inform us about what we are doing to them and what we must start to do about invertebrate animals. Still, it is equally important to, as Birch [28] suggested and Dawkins [35] agreed, invoke the precautionary principle and act ethically towards invertebrates before we know all the details of how they live.

Funding: This research received no external funding.

Institutional Review Board Statement: Not applicable.

Informed Consent Statement: Not applicable.

Conflicts of Interest: The author declares no conflicts of interest.

References

1. McKay, R.; Whitehouse, H. Religion and morality. *Psychol. Bull.* **2015**, *141*, 447–473. [CrossRef]
2. Norenzayan, A. Does religion make people moral? *Behaviour* **2014**, *151*, 365–384. [CrossRef]
3. Singer, P. Ethics and animals: Extending ethics beyond our own species. *Chatauqua J.* **2016**, *1*, 4.
4. Serpell, J.A. Factors influencing human attitudes to animals and their welfare. *Anim. Welf.* **2004**, *13*, S145–S151. [CrossRef]
5. Waldeau, P. Seeing the terrain we walk: Features of the contemporary landscape of "Religion and Animals". In *A Communion of Subjects: Animals in Religion, Science and Ethics*; Columbia U. Press: New York, NY, USA, 2009; pp. 41–61.
6. Szucs, E.; Geers, R.; Jezierski, T.; Sossidou, E.N.; Broom, D.M. Animal welfare in different human cultures, traditions, and religions. *Asian-Aust. J. Anim. Sci.* **2012**, *25*, 1499–1506. [CrossRef]
7. Rahman, M.M.; Yunus, J.; Alenezi, R. Ethical considerations in exploiting, culturing and killing fish towards animal rights in Islam. *IIUM Med. J. Malays.* **2019**, *17*, 287–293. [CrossRef]
8. Finnegan, B. Buddhism and animal ethics. *Philos. Compass* **2017**, *12*, e12424. [CrossRef]
9. Booth, A.L.; Jacobs, H.M. Ties that bind: Native American beliefs as a foundation for environmental consciousness. *Environ. Ethics* **1990**, *12*, 27–43. [CrossRef]
10. Noske, B. Speciesism, anthropocentrism, and non-Western cultures. *Anthrozoos* **1997**, *10*, 183–190. [CrossRef]
11. Frayne, C.T. Animals in Christian and Muslim thought. In *The Routledge Handbook of Religion and Animal Ethics*; Routledge: London, UK, 2018; pp. 201–215.
12. Grandin, T.; Regenstein, J.M. Religious slaughter and animal welfare: A discussion for meat scientists. *Meat Focus Int.* **1994**, *3*, 115–123.
13. Allyn, E.; Parfitt, C. Adult-perpetrated animal abuse: A systematic literature review. *Trauma Violence Abus.* **2019**, *20*, 344–357. [CrossRef] [PubMed]
14. Sinclair, M.; Lee, N.Y.; Hötzel, M.J.; de Luna, M.C.T.; Sharma, A.; Idris, M.; Derkley, T.; Li, C.; Islam, M.A.; Iyasere, O.S.; et al. International perceptions of animals and the importance of their welfare. *Front. Anim. Sci.* **2022**, *3*, 960379. [CrossRef]
15. Rollin, B. Animal welfare across the world. *J. Appl. Anim. Ethics Res.* **2019**, *1*, 146–170. [CrossRef]
16. Noll, S. Broiler chickens and a critique of the epistemic foundation of animal modification. *J. Agric. Environ. Ethics* **2013**, *26*, 273–280. [CrossRef]
17. Knowles, T.G.; Kestin, S.C.; Haslam, S.M.; Brown, S.N.; Green, L.E.; Butterworth, A.; Pope, S.J.; Pfeiffer, D.; Nicol, C.J. Leg disorders in broiler chickens: Prevalence, risk factors, and prevention. *PLoS ONE* **2008**, *3*, e1545. [CrossRef] [PubMed]
18. Edgar, J.L.; Mullan, S.M.; Pritchard, J.C.; McFarlane, U.J.C.; Main, D.C.J. Towards a 'good life' for farm animals: Development of a resource tier framework to achieve positive welfare for laying hens. *Animals* **2013**, *3*, 584–605. [CrossRef]
19. Buller, H.; Roe, E. Modifying and commodifying farm animal welfare: The economization of layer chickens. *J. Rural Stud.* **2014**, *33*, 141–149. [CrossRef]
20. Morton, R.; Hebart, M.L.; Whittaker, A.L. Explaining the gap between the ambitious goals and practical reality of animal welfare law enforcement: A review of the enforcement gap in Australia. *Animals* **2020**, *10*, 482. [CrossRef] [PubMed]
21. Van Zeebroek, S. Karma and Corona: A philosophical perspective on COVID-19 as an outcome of cruelty towards animal by humanity. *Glob. Bioethical Inq.* **2021**, *9*, 5–10.
22. Carnovale, F.; Jin, X.; Arney, D.; Descovich, K.; Guo, W.; Shi, B.; Phillips, C.J.C. Chinese public attitudes towards, and knowledge of, animal welfare. *Animals* **2021**, *11*, 855. [CrossRef]
23. Levine, N.E. A multifaceted interdependence: Tibetan pastoralists and their animals. *Études Mongoles Sibériennes Centrasiatiques Tibétaines* **2019**, *50*. [CrossRef]
24. Fraser, D. Animal ethics and animal welfare science: Bridging the two cultures. *Appl. Anim. Behav. Sci.* **1999**, *65*, 171–189. [CrossRef]
25. Webb, C.E.; Woodford, P.; Huchard, E. Animal ethics and behavioral science: An overdue discussion? *BioScience* **2019**, *68*, 778–788. [CrossRef]
26. Godfrey-Smith, P. *Other Minds: The Octopus and the Evolution of Intelligent Life*; William Collins: Glasglow, UK, 2018.
27. Mikhalevich, I.; Powell, R. Minds without spines: Evolutionarily inclusive animal ethics. *Anim. Sentience* **2020**, *329*, 1. [CrossRef]
28. Birch, J.; Birn, C.; Schnell, A.K.; Browning, H.; Crump, A. *Review of the Evidence of Sentience in Cephalopod Molluscs and Decapod Crustaceans*; Department for Environment, Food & Rural Affairs (DEFRA): London, UK, 2021.
29. Broom, D.M. Science, ethics and public concern about animal welfare. In Proceedings of the 4th European Colloquium on Acute Phase Proteins, Segovia, Spain, 25–26 September 2003; pp. 83–89.
30. Griffin, D.R.; Speck, D.B. New evidence of animal consciousness. *Anim. Cogn.* **2004**, *7*, 15–18. [CrossRef]
31. Brandl, J.L. The puzzle of mirror self-recognition. *Phenomenol. Cogn. Sci.* **2018**, *17*, 279–304. [CrossRef]
32. Pophale, A.; Shimizu, K.; Mano, T.; Iglesias, T.L.; Martin, K.; Hiroi, M.; Asada, K.; Andaluz, P.G.; Van Dinh, T.T.; Meshulam, L.; et al. Wake-like skin patterning and neural activity during octopus sleep. *Nature* **2023**, *619*, 129–134. [CrossRef]
33. Carruthers, P. The problem of animal consciousness. In *Proceedings and Addresses of the American Philosophical Association*; American Philosophical Association: San Diego, CA, USA, 2018; Volume 92, pp. 179–204.
34. Baars, B.J. Global workspace theory of consciousness: Toward a cognitive neuroscience of human experience. *Prog. Brain Res.* **2005**, *150*, 45–53.

35. Dawkins, M. Animal welfare and the paradox of animal consciousness. In *Why Animals Matter—Animal Consciousness, Animal Welfare and Human Well-Being*; Oxford U. Press: Oxford, UK, 2014.
36. Elwood, R.W. Behavioural indictors of pain and suffering in Arthropods and might pain bite back? *Animals* **2023**, *13*, 2602. [CrossRef]
37. Sneddon, L.U.; Elwood, R.W.; Adamo, S.; Leach, M.C. Defining and assessing animal pain. *Anim. Behav.* **2014**, *97*, 201–212. [CrossRef]
38. Crook, R.J. Behavioral and neurophysiological evidence suggests affective pain experience. *iScience* **2021**, *24*, 102229. [CrossRef] [PubMed]
39. Bateson, M.; Desire, S.; Gartside, S.E.; Wright, G.A. Agitated honeybees exhibit pessimistic cognitive bias. *Curr. Biol.* **2011**, *21*, 1070–1073. [CrossRef]
40. Hampton, J.O.; Hyndman, T.H.; Allen, B.L.; Fisher, B. Animal harms and food production: Informing ethical choices. *Animals* **2021**, *11*, 1225. [CrossRef]
41. Clarke, S.; Savalescu, J. Rethinking our assumptions about moral status. In *Rethinking Moral Status*; Clarke, S., Zohny, H., Savalescu, J., Eds.; Oxford U Press: Oxford, UK, 2021; pp. 1–12.
42. DeGrazia, D. Sentience and consciousness as bases for attributing interests and moral status: Considering the evidence and speculating slightly beyond. In *Neuroethics and Nonhuman Animals*; Johnson, L.S., Fenton, A., Shriver, A., Eds.; Springer: Berlin/Heidelberg, Germany, 2020; pp. 17–32.
43. Roth, G. Convergent evolution of complex brains and high intelligence. *Philos. Trans. R. Soc. B Biol. Sci.* **2015**, *370*, 20150049. [CrossRef] [PubMed]
44. Pennartz, C.M.A.; Farisco, M.; Evers, K. Indicators and criteria of consciousness in animals and intelligent machines: An inside-out approach. *Front. Syst. Neurosci.* **2019**, *13*, 25. [CrossRef] [PubMed]
45. Mather, J.A.; Andrade, M. Can we use introspection to assist the study of decision making and understand consciousness in cephalopods? Symposium on Introspection. *J. Conscious. Stud.* **2023**, *9–10*, 164–173. [CrossRef]
46. Barrett, M.; Adcock, S.J.J. Animal welfare science: An integral piece of sustainable insect agriculture. *J. Insects Food Feed* **2023**, *1*, 1–15. [CrossRef]
47. Kneiss, J. Bentham on animal welfare. *Br. J. Hist. Philos.* **2019**, *27*, 556–572. [CrossRef]

animals

MDPI

Article

Effects of Acetic Acid and Morphine in Shore Crabs, *Carcinus maenas:* Implications for the Possibility of Pain in Decapods

Stuart Barr and Robert W. Elwood *

School of Biological Sciences, Queen's University, Belfast BT9 5DL, UK; s_barr@hotmail.com
* Correspondence: r.elwood@qub.ac.uk

Simple Summary: Injection of acetic acid into one leg of a crab caused rubbing of that leg and holding the leg off the floor of a tank. Such activities directed towards the site of a noxious stimulus are consistent with the idea of pain. Some crabs also cast off the leg injected with acid. Because that occurred in association with possible pain-related behaviour, it too might be caused by pain. Prior morphine injection caused various behavioural changes, but morphine did not ameliorate the responses to acetic acid. Therefore, morphine does not act as an analgesic, and this agrees with previous work. Nevertheless, the directed behaviour that follows injection into a leg agrees with other studies and provides additional evidence suggesting that these animals experience pain.

Abstract: Noxious chemicals, coupled with morphine treatment, are often used in studies on pain in vertebrates. Here we show that injection of morphine caused several behavioural changes in the crab, *Carcinus maenas*, including reduced pressing against the sides of the enclosure and more rubbing and picking at the mouth parts and, at least for a short time, more defensive displays. Subsequent injection of acetic acid into one rear leg caused rubbing of the injected leg and the injected leg was held vertically off the ground. These activities directed at or involving the specific leg are consistent with previous observations of directed behaviour following noxious stimuli and are consistent with the idea that decapods experience pain. Further, acetic acid but not injection of water induced autotomy of the injected leg in these animals. Because autotomy is temporally associated with directed behaviour, it is possible that the autotomy is a pain-related response. Acetic acid is clearly a noxious substance when applied to decapods. However, morphine had no effect on the activities associated with acetic acid injection and thus there is no evidence for an analgesic effect. Further, the injection of acetic acid did not interfere with behavioural effects of morphine. The activities directed towards the site of injection are like those observed with injection, or with external application, of various noxious substances and the present study adds to a growing body of knowledge about possible pain in decapods.

Keywords: autotomy; pain; acetic acid; morphine; *Carcinus maenas*; decapod

1. Introduction

Studies on mammalian pain frequently employ noxious chemicals including acetic acid and formalin, to elicit marked behavioural changes that direct activities towards the site of injection, or other, more extreme, responses, such as writhing in mice, *Mus musculus*, that are interpreted as the animal being in pain [1,2]. The behaviour of these animals can be compared to others that are treated with potential analgesics to determine if those, presumably, pain responses are reduced [1,2].

This approach was taken in studies on fish to investigate if that taxon could experience pain [3]. The lips of trout were injected with either acetic acid or saline control, and the former rubbed their lips on the substrate and showed abnormal rocking, which was not seen in controls. However, morphine injection reduced the rubbing and rocking. This experiment provided ideas for an early study on potential pain experience in decapod crustaceans [4,5]. That study used glass prawns (*Palaemon elegans*), which had an antenna brushed with a local anaesthetic (benzocaine) or saline control and then later had the same antenna brushed with acetic acid or caustic soda or saline [5]. Without the local anaesthetic, the noxious chemicals caused a marked increase in rubbing the antenna against the walls of the tank, and grooming directed at the treated antenna, but these responses were reduced in those pretreated with benzocaine [5]. However, another study found no increase in behaviour directed to the site of application in three species of decapods following brushing of hydrochloric acid or caustic soda to an antenna [6]. That study concluded that decapods were not affected by chemicals of high or low pH. Several following studies questioned that conclusion. Shore crabs, *Carcinus maenas*, showed directed behaviour when their mouthparts or an eye was brushed with acetic acid [7], and glass prawns (*P. elegans*) used their chelipeds to rub an eye if the eye had been brushed with caustic soda [8]. Crabs (*Hemigrapsus sanguineus*) responded when formalin, another substance that causes pain in humans, was injected into a claw and the animal reduced the use of that appendage when walking and pressed that claw against its carapace [9]. Crabs also shook and rubbed the injected claw, and some autotomized the injected claw, but none of these activities were seen in saline-injected controls [9]. One aim of the present study is to ask if acetic acid, injected into a walking leg, induces behaviour that is directed at the injected leg. We also recorded autotomy, as that too might be mediated by pain [10]. This should answer the question of whether acetic acid can influence behaviour in a way that is consistent with the idea of pain in decapods. It also provides a novel method of assessing the effects of acetic acid because previous tests employed external application, whereas this is by injection.

While morphine has been shown to be a powerful analgesic in mammals, and reduces responses to various noxious stimuli, it may also have other, non-analgesic effects [11,12]. That is, morphine may cause changes in behaviour, and it is possible that these may mask, or be mistaken for, analgesia. This appeared to be the case when testing for analgesic effects of morphine in crustaceans. For example, responses to an electrical shock were reduced by dose-related morphine treatment in the mantis shrimp, *Squilla mantis* [13], and in the crab, *Chasmagnathus granulatus* [14]. In the latter, this apparent analgesia was reversed by the opioid antagonist, naloxone [14]. However, escape responses to a moving shadow *C. granulatus* were also reduced by morphine injection [15]. In this case, the behavioural effect of morphine cannot be attributed to analgesia. However, the effect could be due to a general reduction in perception or ability to move, and thus the proposed analgesic effect of morphine was questioned [15]. To test this idea, an experiment was devised in which crabs were repeatedly offered the choice to move to a dark shelter to escape a brightly lit area [16]. Some crabs were shocked within the shelter and some crabs had prior morphine treatment. The idea was that if morphine had an analgesic effect, then morphine-treated crabs would use the shelter even though they received a shock. It was found that crabs given morphine did not use the shelter in early trials, irrespective of being shocked or not, but did so in later trials. The conclusion was that morphine offered no analgesic effect for electric shock but did alter the general responsiveness or ability to move [16]. The present experiment further investigates the effects of morphine on the behavioural responses of shore crabs, *Carcinus maenas*, and tests for analgesic effects following acetic acid injection to a walking leg. It also tests for behaviour following acetic acid injection that is directed at the injected leg, which, if found, would be consistent with the idea of pain in crabs. In addition, it examines if autotomy occurs following acetic acid injection, as was noted with formalin [9], and if that reaction might be reduced by morphine.

2. Materials and Methods

2.1. Animal Collection and Care

European shore crabs (*Carcinus maenas*, carapace width = 5–8 cm) were collected from Bar Hall Bay, Strangford Lough (OS; J 617464), between May and July 2008, using baited pots. Only fully intact crabs were transported to Queen's University, Belfast. They were housed in plastic tanks (76 × 38 × 17 cm) filled with aerated natural seawater, to a depth of 3 cm, and seaweed for shelter (*Ascophyllum nodosum*), in a cold room maintained at 11–13 °C with a twelve-hour light/dark regime. The crabs were fed fresh mackerel pieces every 2 days, after which the seawater was changed. Before performing the experiment, carapace width was measured and health status was noted once again as, during the storage time, some crabs were damaged due to conflicts. Only healthy, intact crabs were randomly assigned (by drawing tokens from a bag) to eight experimental groups (N = 12 per group), each of which involved two treatments and two observation periods in a 2 × 4 design. After use, the crabs were returned to the cold room and housed in new tanks, before being released back in Strangford Lough.

2.2. Experimental Treatments

2.2.1. First Treatment

The first treatment was either a 100 μg/g injection of pharmaceutical-grade morphine sulphate solution (60 mg/2 mL), calculated by first weighing the animal to determine the correct volume, or an equivalent amount of distilled water. This dose was the same as that employed in previous studies [14–16]. This first injection was randomly chosen and administered in the cephalothoracic–abdominal membrane, using 0.3 mL insulin syringes, while the animal was briefly held on a treatment table. Following first treatment, the animal was placed into the observation tank (30 × 19 × 20 cm) containing fresh natural seawater (11–13 °C) to a depth of 1 cm. The observation tank was housed in an observation chamber (illuminated by a 60 W light bulb) behind a one-way mirror and behaviour recorded for 5 min.

Contact with the sides of the tank was recorded. This involved the crab pressing into the sides of the tank with its chelipeds, backing into the tank sides or in some cases attempting to climb the wall of the tank. The amount of time spent rubbing and picking the mouthparts with the chelipeds and the time spent in the lateral merus display (elevated carapace and extended chelipeds) was also recorded.

2.2.2. Second Treatment

The crab was removed from the observation tank for the second treatment and again briefly held on the treatment table. The second injection was 0.2 mL of distilled water, 0.2 mL of 0.25% acetic acid, 0.2 mL of 0.5% acetic acid or 0.2 mL of 1.0% acetic acid. These concentrations of acetic acid were selected after preliminary trials. The treatment was randomly chosen and administered to the carpus/propus joint of a randomly chosen hind leg. This joint is about halfway along the length of the leg. Following treatment, the crab was placed into the observation tank again and behaviour recorded for 5 min. The same activities as before were recorded. In addition, rubbing of the injected leg with adjacent limbs, vertical extension of the injected limb, and incidence and timing of leg autotomy were recorded. Because morphine effects last for about 20 min [16], both observation periods were completed before the effects declined.

2.3. Statistical Methods

2.3.1. First Observation Period

Effects of morphine vs. water injections on durations of contact with the sides of the tank, mouth rubbing and picking with the chelipeds, and defensive reaction were transformed to log(x + 1) and analysed using *t*-tests. The transformation improved the normal distribution of the data. The number of crabs showing defensive display responses was analysed using χ^2 tests.

2.3.2. Second Observation Period

The incidence of rubbing the injected leg, holding the injected leg vertically, and autotomy was analysed using χ^2 tests for effect of first treatment (morphine or distilled water) and for effect of second treatment (distilled water or acetic acid). The durations of contact with the sides of the tank and of mouth rubbing and picking with the chelipeds were transformed to log(x + 1) and analysed using a 2-factor ANOVA (factor 1—morphine or distilled water; factor 2—distilled water, 0.25%, 0.5% and 1.0% acetic acid).

2.4. Ethical Considerations

This experiment was conducted in 2008, when there was little or no support for the idea of pain in decapod crustaceans. There were no legal restrictions on experiments and no requirement for ethical review on this group of animals within the United Kingdom. Nevertheless, we kept the numbers of animals low in each experimental group ($n = 12$ per group) and we kept the level of noxious stimulus as low as possible. All crabs recovered and were returned to the shore. Autotomy of a single leg does not impede locomotion and is not expected to interfere with feeding and mating. It thus poses less of a cost for the crab than does loss of a cheliped [17]. Further, the crabs were not fully grown, and autotomized walking legs should regenerate when the animal moults and grows. The results of similar experiments have changed the legal situation for decapods within the United Kingdom, which now recognizes that these animals are sentient. Despite this, there has been no change in the UK legal requirements for research on these animals and thus the experiment is fully compliant with current UK regulations. Nevertheless, we suggest that researchers should take the potential sentience of these animals into account when designing future studies rather than waiting for legal change and refer to guidelines on the use of wild animals [18] and, specifically, crustaceans [19].

3. Results

3.1. First Observation Period

Crabs that received morphine pressed against the sides of the tank for significantly less time ($t_{94} = 2.92$, $p < 0.01$; Figure 1), but rubbed and picked their mouthparts with their chelipeds significantly more ($t_{94} = 8.96$, $p < 0.0001$; Figure 2), and were more likely to show defensive displays than were those injected with water (20/48 for morphine and 2/48 for water, $\chi^2_1 = 17.04$, $p < 0.0001$).

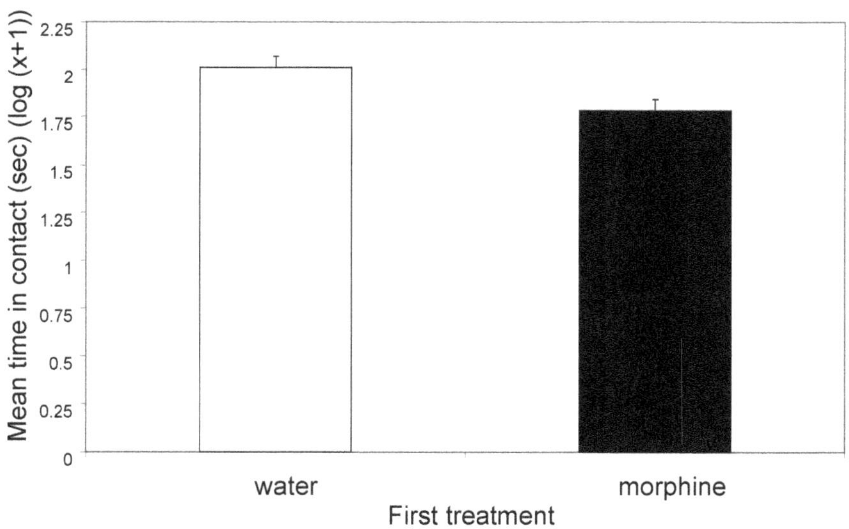

Figure 1. The mean (±SE) time (s) spent in contact with the sides of the tank in the first observation (log (x + 1)). The means equate to 100 s (water) and 60 s (morphine).

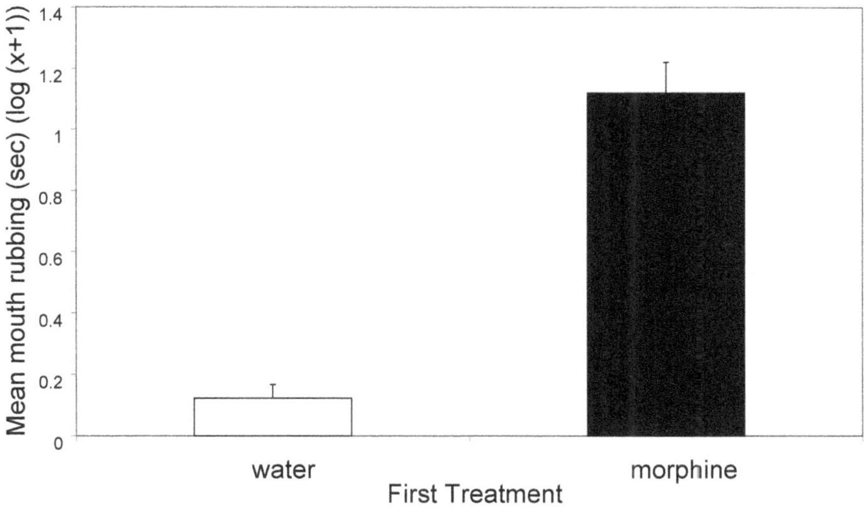

Figure 2. The mean (±SE) time (s) spent rubbing and picking the mouthparts with the chelipeds in the first observation (log(x + 1)). The means equate to 1.5 s (water) and 12.5 s (morphine).

3.2. Second Observation Period

Contact with the sides of the tank in the second period was significantly affected by first treatment, with less contact recorded when the first treatment was morphine ($F_{1, 88} = 43.99$, $p < 0.0001$; Figure 3). Contact with the sides of the tank was not significantly affected by second treatment ($F_{3, 88} = 0.861$, $p = 0.46$) and there was no interaction between first and second treatments ($F_{3, 88} = 1.82$, $p = 0.15$).

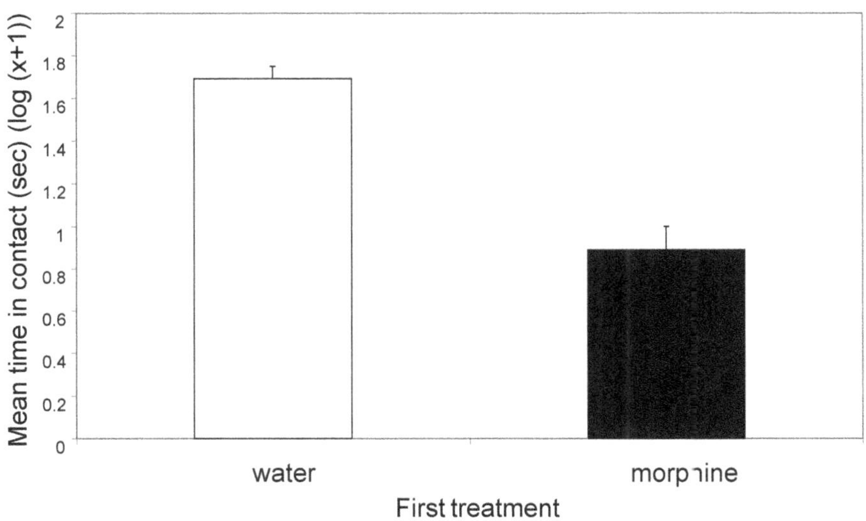

Figure 3. The mean (±SE) time (s) spent in contact with the sides of the tank in the second observation (log(x + 1)). The means equate to 50 s (water) and 8 s (morphine).

There was a significant effect of first treatment on the amount of mouth rubbing during the second observation, with those crabs that received morphine showing more rubbing than those that received distilled water ($F_{1, 88} = 50.25$, $p < 0.0001$; Figure 4). There was no effect of second treatment on the amount of mouth rubbing ($F_{3,88} = 1.74$, $p = 0.165$), and there was no interaction between first and second treatments ($F_{3,88} = 1.53$, $p = 0.212$).

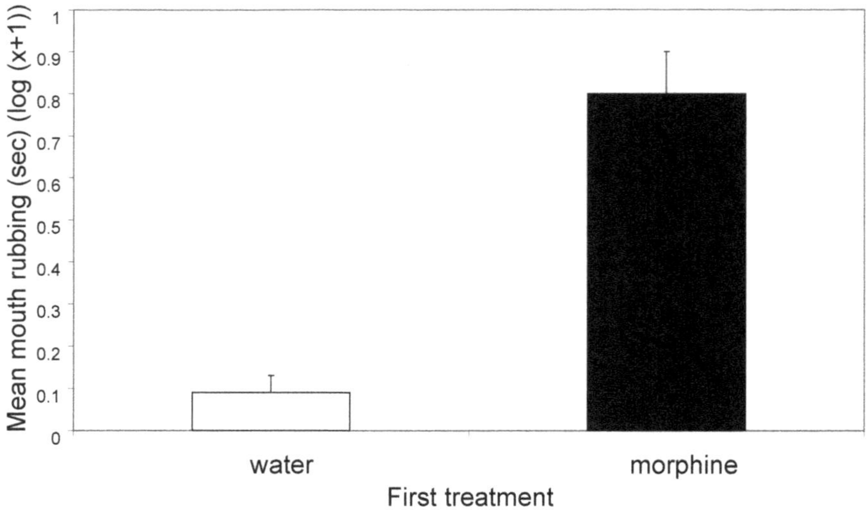

Figure 4. The mean (±SE) time (s) spent rubbing and picking the mouthparts with the chelipeds in the second observation (log(x + 1)). The means equate to 1.2 s (water) and 6.5 s (morphine).

There were only six cases of crabs showing the defensive response during the second period, and this sample was too small for any meaningful analysis.

The occurrence of rubbing the injected leg varied significantly between concentrations of acid (0/24 for distilled water, 10/24 for 0.25% acetic acid, 7/24 for 0.5% acetic acid, 8/24 for 1.0% acetic acid; $\chi^2_3 = 12.28$, $p = 0.0065$), but rubbing was not influenced by the first treatments of morphine or distilled water (15/48 versus 10/48; $\chi^2_1 = 0.35$, $p = 0.24$). Vertical extension of the injected leg during the second observation also varied significantly with concentration of acid (0/24 for distilled water; 5/24 for 0.25% acetic acid, 9/24 for 0.5% acetic acid, 10/24 for 1.0% acetic acid; $\chi^2_3 = 13.78$, $p = 0.0032$), but was not affected by the first treatment of distilled water or morphine (12/24 versus 12/24; $\chi^2_1 = 0.000$, $p = 1.0$). The incidence of autotomy also varied significantly between concentrations of acid (0/24 for distilled water, 2/24 for 0.25% acetic acid, 20/24 for 0.5% acetic acid, 20/24 for 1.0% acetic acid; $\chi^2_3 = 61.46$, $p = 0.0079$), but was not influenced by the first treatment of morphine or distilled water (23/48 versus 19/48; $\chi^2_1 = 0.677$, $p = 0.41$). Sometimes the autotomy occurred before the animal could be returned to the observation tank (fast), but for others it occurred after the animal was returned to the observation tank (slow). There was no effect of morphine treatment on the speed of autotomy (16 fast; 7 slow for water treatment, 10 fast and 9 slow for morphine treatment $\chi^2_1 = 0.65$, $p = 0.42$).

4. Discussion

4.1. Effects of Morphine on Behaviour

The first treatment with morphine or distilled water had marked effects on the behaviour of the animals in the first and the second observation periods. Morphine treatment caused animals to spend less time in contact with the walls of the tank in period one (Figure 1) and in period two (Figure 3). Further, morphine elicited more rubbing and picking at the mouthparts in period one (Figure 2) and in period two (Figure 4). The two observation periods were 5 min each and about 1 min was required for the injection into a leg prior to the second periods, so these observations all took place within 11 min of the morphine or control injections.

Effects on the behaviour of decapods following morphine administration have been noted in previous studies. Shore crabs, *C. maenas*, injected with morphine reduced the use of dark shelters to escape from bright light [16] for about 20 min. This fits with the present observation of reduced seeking of shelter as seen by less contact with the walls of the tank. Reduced seeking of shelter has been explained as a reduced ability to walk [16] but it could be a reduced motivation to seek shelter. Morphine injection reduced the escape response to a moving shadow in *C. granulatus*, which might have been due to a general reduction in perception or ability to move [15]. Further, the picking and rubbing of mouth parts following morphine injection was like observations on crayfish, *Orconectes rusticus*, which showed stereotypic behaviour, including grooming, tail flipping, mouthpart movements, repetitive actions, and mild tremor [20].

In the present study, morphine caused more animals to show the defensive display in period one, but not in period two, when very few animals showed these displays. Morphine seems to reduce shelter use and replace it with defensive displays. However, the increase in protective displays was short-lived. This could be because the effects of morphine declined rapidly, or it could be due to a rapid habituation to the novel environment. Regardless, we see that morphine has various effects on the behaviour of decapods, but it is not clear how morphine works on the nervous system of decapods to cause those effects [21].

4.2. *Effects of Acetic Acid*

Injection of acetic acid, of any dilution, into a joint of a hind leg caused marked changes in behaviour. The crabs were more likely to rub the injected leg with other legs and to hold the injected leg vertically. Because these activities were not seen when distilled water was injected, they cannot be caused by the minor tissue damage from the injection; rather, the acetic acid caused the effect. Acetic acid has long been used in studies on possible pain in vertebrates [3] and crustaceans [5,7], and behaviour directed at the site of noxious stimulation has been taken to be consistent with the idea of pain experience [22–24]. Those previous studies on crustaceans, however, involved external application by brushing acetic acid solution onto an antenna, an eye or mouth parts. The present study involved injection of acetic acid and thus was more like a study on trout that showed elevated rubbing and abnormal activities when the lip of the fish was injected [3]. It was also like a study on the shore crab, *H. sanguineus*, in which claw-bearing limbs were injected with formalin, and shaking, rubbing and unusual postures of that limb occurred [9]. These studies have demonstrated that both acetic acid and formalin causes decapods to show distinct activities that are consistent with the idea of pain, and similar effects have been noted with caustic soda brushed onto an eye in glass prawns [8].

Acetic acid injection also caused animals to autotomize the injected leg, and this occurred primarily with the higher concentrations of acetic acid. Sometimes this autotomy occurred almost immediately following acetic acid injection, i.e., before observations could commence for that period, and this obviously precluded holding the leg vertically and/or rubbing the leg. However, when autotomy occurred after the animal was placed back into the observation chamber, rubbing and holding the leg vertically often preceded autotomy. Rubbing and holding the leg vertically could also occur without autotomy and this was particularly the case with the lowest concentration of acetic acid. It seems that low concentrations of acetic acid induced leg-directed activities, but a stronger solution was usually required for autotomy. In the present study, animals were observed to rub the site of autotomy for prolonged periods (60 s), which is consistent with the idea of pain. Autotomy was also seen in *H. sanguineus* when the clawed appendage was injected with formalin [9], and, like the present study, that was often preceded by shaking and rubbing the appendage. Autotomy is a well-known response of decapods to noxious stimuli to an appendage and occurs in response to being held [25], heat [26], electric shock [16], a small cut at a joint [27], injection of dopamine [28], as well injection with formalin [9] and injection with acetic acid reported here. There has been speculation that autotomy in arthropods is mediated, at least in some cases, by a pain-like state [10,22,29]. Spiders, *Argiope* spp., for example, will autotomize a leg when injected with chemicals known to cause pain in humans but not others that do not cause pain in humans [30]. On the other hand, there is no evidence to suggest that autotomy due to being held by an appendage is mediated by pain; rather, it seems to be a decision to pay a cost to escape from a potential predator. The present results, however, are suggestive of pain-like mediation because autotomy is often preceded by rubbing and holding in an unusual posture, which are activities that fit a criterion of pain.

Autotomy in decapods occurs at the joint between the coxa and the basis (basi-ishiopodite). It is caused by contraction of muscles normally used in locomotion and fractures a specific plane that heals rapidly [31,32]. The muscles are activated by motor neurons, and thus, autotomy seems to follow central decision making [32]. It is thought that autotomy following dopamine injection is due to the dopamine acting at the site normally controlled by motor neurons [28]. Autotomy appears to vary with other motivations and the decision to autotomize may include information about costs and benefits of losing the limb [33]. Thus, it should be amenable to experiments that examine trade-offs, a key criterion of pain. Because there is a cost to autotomy, coupled with the cost of subsequent regeneration of the appendage [25], then it would be expected to be induced following a decision-making process that maximizes the fitness pay-off. This could be a better outcome than that achieved by a standard reflex response [34].

4.3. Effects of Morphine on Responses to Acetic Acid

Although morphine influenced the behaviour of crabs, there is no indication that it affected responses to the injection of acetic acid. That is, it did not reduce rubbing of the injected appendage or holding the appendage in a vertical position. These activities directed at the site of a noxious stimulus have the appearance of being consistent with pain, but there is no analgesic effect of morphine. Further, there was no effect of morphine on the probability of autotomy or the speed of autotomy. This might be due to an absence of any pain-like feeling, or it could be that the physiology of crabs is very different from that of vertebrates with respect to responses to morphine. Morphine has analgesic effects in vertebrates [1–3], but the evidence for such effects in crustaceans is weak or lacking [35]. However, local anaesthetics, such as benzocaine, are effective in reducing responses to noxious stimuli in decapods [5,8]. Benzocaine blocks the transmission of action potentials along axons, whereas opiates block transmission across synapses by inhibiting specific neurotransmitters [35].

Studies on mantis shrimps, *S. mantis*, and the crab, *C. granulatus*, have shown reduced responses to electric shock following morphine injection [13,14], and this was reversed by the opioid antagonist, naloxone [14]. However, morphine also reduced the responsiveness to visual stimuli [15] and thus there was the possibility that morphine had effects on locomotor or perceptual abilities. One experiment on shore crabs that had the expectation of greater movement into a dark chamber if there was an analgesic effect of morphine found no analgesic effect. Rather, morphine had general effects on behaviour that reduced mobility, and perhaps the effects of visual stimuli [16]. We have shown in the present study that morphine can alter behaviour, but there is no indication that it provided analgesia, because of the lack of effect on any behaviour directed to or involving the injected leg.

5. Conclusions

Crabs show behavioural changes to both morphine and acetic acid, but these are independent of each other. That is, morphine does not influence the responses to acetic acid and acetic acid does not modify the responses to morphine. However, the responses to acetic acid are directed at the specific limb that was injected, and directed behaviour has been accepted as a criterion for pain [22–24,36]. This measure is used extensively as part of the formalin test in pharmaceutical investigations [37] in which the level of activities directed at the site of injection in a leg is used as a measure of pain in rodents. Further, autotomy of the limb injected with a noxious chemical, but not with distilled water, has been suggested as an indicator of pain [10,22,29]. Thus, our present findings are consistent with the idea of pain in decapods. However, pain is impossible to prove in non-verbal species and we do not claim to demonstrate pain with these findings. Nevertheless, there is a body of studies on potential pain in decapods that test various criteria for pain, and these too have fulfilled those criteria and thus are also consistent with the idea of pain. For example, we see swift avoidance learning [38–40]. There are trade-offs between avoidance of a noxious stimulus and other motivational requirements, such as hermit crabs, *Pagurus bernhardus*, accepting electric shock to keep high-quality shells [41,42], or to avoid predators [43]. Further, shore crabs accept electric shocks to avoid bright but not dull light [39]. Following electric shocks, decapods show signs of anxiety [40,44,45] and show physiological stress responses after electric shocks [44–46] or injury [27]. Hermit crabs that are shocked within their shells show long-term reduction in their motivation to retain that shell (24 h) [47]. It is the accumulation of studies that are consistent with the idea of pain in decapods that has led to an increasing acceptance of the possibility of pain in this taxon [24]. This increasing acceptance is exemplified by the policy statement by the British Veterinary Association (2020) [48] and the recent inclusion of decapods in the Animal Welfare (Sentience) Act (2022) of the United Kingdom [49]. Although this act recognizes decapods as sentient, it has not yet resulted in changes in legal requirements to improve their welfare in scientific research or in aquaculture, fishing, transport or use in food preparation within the United Kingdom. Finally, we recognize that decapods are a diverse group and that indicators, suggestive of pain, have not been investigated in all of the major groups [24,50].

Author Contributions: Conceptualization, S.B. and R.W.E.; methodology, S.B. and R.W.E.; formal analysis, S.B. and R.W.E.; investigation, S.B.; resources, R.W.E.; data curation, S.B. and R.W.E.; writing—original draft preparation, S.B. and R.W.E.; writing—review and editing, R.W.E.; supervision, R.W.E.; project administration, R.W.E.; funding acquisition, S.B. and R.W.E. All authors have read and agreed to the published version of the manuscript.

Funding: This research was funded by Department of Agriculture and Rural Development of Northern Ireland and Queen's University, Belfast.

Institutional Review Board Statement: Not applicable.

Informed Consent Statement: Not applicable.

Data Availability Statement: Data are contained within the article.

Acknowledgments: We thank Gillian Riddell for technical support, and two anonymous reviewers for some good suggestions that improved the paper.

Conflicts of Interest: The authors declare no conflicts of interest.

References

1. Gregory, N.S.; Harris, A.L.; Robinson, C.R.; Dougherty, P.M.; Fuchs, P.N.; Sluka, K.A. An overview of animal models of pain: Disease models and outcome measures. *J. Pain* **2013**, *14*, 1255–1269. [CrossRef] [PubMed]
2. Lee, G.; Park, J.; Kim, M.S.; Seol, G.H.; Min, S.S. Analgesic effects of eucalyptus essentialoil in mice. *Korean J. Pain* **2019**, *32*, 79–86. [CrossRef]
3. Sneddon, L.U. The evidence for pain in fish: The use of morphine as an analgesic. *Appl. Anim. Behav. Sci.* **2003**, *83*, 153–162. [CrossRef]
4. Elwood, R.W. Potential pain in fish and decapods: Similar experimental approaches and similar results. *Front. Vet. Sci.* **2021**, *8*, 631151. [CrossRef]
5. Barr, S.; Laming, P.R.; Dick, J.T.; Elwood, R.W. Nociception or pain in a decapod crustacean? *Anim. Behav.* **2008**, *75*, 745–751. [CrossRef]
6. Puri, S.; Faulkes, Z. Do decapod crustaceans have nociceptors for extreme pH? *PLoS ONE* **2010**, *5*, e10244. [CrossRef]
7. Elwood, R.W.; Dalton, N.; Riddell, G. Aversive responses by shore crabs to acetic acid but not to capsaicin. *Behav. Process.* **2017**, *140*, 1–5. [CrossRef] [PubMed]
8. Barr, S.; Elwood, R.W. The Effects of Caustic Soda and Benzocaine on Directed Grooming to the Eyestalk in the Glass Prawn, *Palaemon elegans*, Are Consistent with the Idea of Pain in Decapods. *Animals* **2024**, *14*, 364. [CrossRef] [PubMed]
9. Dyuizen, I.V.; Kotsyuba, E.P.; Lamash, N.E. Changes in the nitric oxide system in the shore crab *Hemigrapsus sanguineus* (Crustacea, decapoda) CNS induced by a nociceptive stimulus. *J. Exp. Biol.* **2012**, *215*, 2668–2676. [CrossRef]

10. Sherwin, C.M. Can invertebrates suffer? Or how robust is argument-by-analogy? *Anim. Welf.* **2001**, *10*, 103–108. [CrossRef]
11. Oliverio, A.; Castellano, C.; Puglisi-Allegra, S. Psychobiology of Opioids. *Int. Rev. Neurobiol.* **1984**, *25*, 277–337. [CrossRef] [PubMed]
12. D'Amato, F.R.; Castellano, C. Behavioral effects of morphine in mice: Role of experimental housing. *Pharmacol. Biochem. Behav.* **1989**, *34*, 361–365. [CrossRef] [PubMed]
13. Maldonaldo, H.; Miralto, A. Effect of morphine and naloxone on a defensive response of the mantis shrimp (*Squilla mantis*). *J. Comp. Physiol.* **1982**, *147*, 455–459. [CrossRef]
14. Lozada, M.; Romano, A.; Maldonado, H. Effect of morphine and naloxone on a defensive response of the crab *Chasmagnathus granulatus*. *Pharm. Biochem. Behav.* **1988**, *30*, 635–640. [CrossRef] [PubMed]
15. Tomsic, T.M.V.; Maldonado, H. Habituation to a danger stimulus in two semiterrestrial crabs: Ontogenic, ecological and opioid modulation correlates. *J. Comp. Physiol. A* **1993**, *173*, 621–633. [CrossRef]
16. Barr, S.; Elwood, R.W. No evidence of morphine analgesia to noxious shock in the shore crab, *Carcinus maenas*. *Behav. Process.* **2011**, *86*, 340–344. [CrossRef] [PubMed]
17. Patterson, L.; Dick, J.T.A.; Elwood, R.W. Claw loss and feeding ability in the edible crab, *Cancer pagurus*: Implications of fishery practice. *Appl. Anim. Behav. Sci.* **2009**, *116*, 302–305. [CrossRef]
18. Soulsbury, C.; Bateson, M.; Braithwaite, V.; Cotter, S.; Elwood, R.W.; Whiting, M.; Wilkinson, A.; Collins, L.M. The welfare and ethics of research involving wild animals: A primer. *Methods Ecol. Evol.* **2020**, *11*, 1164–1181. [CrossRef]
19. Crump, A.; Fischer, B.; Arnott, G.; Birch, J.; Briffa, M.; Browning, H.; Coates, C.; Elwood, R.; Khan, N.; Thakrar, U.; et al. *Guidelines for Protecting and Promoting Decapod Crustacean Welfare in Research*; Insect Welfare Research Society: Indianapolis, IN, USA, 2024. Available online: https://www.insectwelfare.com/research-guidelines (accessed on 14 February 2024).
20. Nathaniel, T.I.; Panksepp, J.; Huber, R. Effects of a single and repeated morphine treatment on conditioned and unconditioned behavioral sensitization in crayfish. *Behav. Brain Res.* **2010**, *207*, 310–320. [CrossRef]
21. de Souza Valente, C. Anaesthesia of decapod crustaceans. *Vet. Anim. Sci.* **2022**, *16*, 100252. [CrossRef]
22. Elwood, R.W. Pain and Suffering in Invertebrates? *ILAR J.* **2011**, *52*, 175–184. [CrossRef] [PubMed]
23. Elwood, R.W. Discrimination between nociceptive reflexes and more complex responses consistent with pain in crustaceans. *Philos. Trans. R. Soc. B Biol. Sci.* **2019**, *374*, 20190368. [CrossRef] [PubMed]
24. Crump, A.; Browning, H.; Schnell, A.; Burn, C.; Birch, J. Sentience in decapod crustaceans: A general framework and review of the evidence. *Anim. Sentience* **2022**, *7*, 1. [CrossRef]
25. Maginnis, T.L. The costs of autotomy and regeneration in animals: A review and framework for future research. *Behav. Ecol.* **2006**, *17*, 857–872. [CrossRef]
26. Fiorito, G. Is there 'pain' in invertebrates? *Behav. Process.* **1986**, *12*, 383–388. [CrossRef] [PubMed]
27. Patterson, L.; Dick, J.T.A.; Elwood, R.W. Physiological stress responses in the edible crab, *Cancer pagurus* to the fishery practice of de-clawing. *Mar. Biol.* **2007**, *152*, 265–272. [CrossRef]
28. Sy, C.; Airriess, C.N. Dopamine–Stimulated Limb Autotomy in the Dungeness Crab, *Cancer magister*. *J. Crust. Biol.* **2002**, *22*, 697–703. [CrossRef]
29. Elwood, R.W. Behavioural Indicators of Pain and Suffering in Arthropods and Might Pain Bite Back? *Animals* **2023**, *13*, 2602. [CrossRef]
30. Eisner, T.; Camazine, S. Spider leg autotomy induced by prey venom injection: An adaptive response to "pain"? *Proc. Natl. Acad. Sci. USA* **1983**, *80*, 3382–3385. [CrossRef]
31. McVean, A. The nervous control of autotomy in *Carcinus maenas*. *J. Exp. Biol.* **1974**, *60*, 432–436. [CrossRef] [PubMed]
32. Findlay, I.; McVean, A.R. The nervous control of limb autotomy in the hermit crab *Pagurus bernhardus* (L.) and the role of the cuticular stress detector, CSD1. *J. Exp. Biol.* **1977**, *70*, 93–104. [CrossRef]
33. Darnell, M.Z.; Patricia, R.Y.; Backwell, P.R.Y.; Munguia, P. Frequency and latency of autotomy of a sexually selected fiddler crab appendage. *J. Exp. Mar. Biol. Ecol.* **2020**, *523*, 151255. [CrossRef]
34. Farnsworth, K.D.; Elwood, R.W. Why it hurts: With freedom comes the biological need for pain. *Anim. Cogn.* **2023**, *26*, 1259–1275. [CrossRef]
35. Rotllant, G.; Llonch, P.; García del Arco, J.A.; Chic, Ò.; Flecknell, P.; Sneddon, L.U. Methods to Induce Analgesia and Anesthesia in Crustaceans: A Supportive Decision Tool. *Biology* **2023**, *12*, 387. [CrossRef]
36. Weary, D.M.; Neil, L.; Flower, F.C.; Fraser, D. Identifying and preventing pain in animals. *Appl. Anim. Behav. Sci.* **2006**, *100*, 64–76. [CrossRef]
37. Munro, G. Pharmacological assessment of the rat formalin test utilizing the clinically used analgesic drugs gabapentin, lamotrigine, morphine, duloxetine, tramadol and ibuprofen: Influence of low and high formalin concentrations. *Eur. J. Pharmacol.* **2009**, *605*, 95–102. [CrossRef] [PubMed]
38. Magee, B.; Elwood, R.W. Shock avoidance by discrimination learning in the shore crab (*Carcinus maenas*) is consistent with a key criterion for pain. *J. Exp. Biol.* **2013**, *216*, 353–358. [CrossRef]
39. Barr, S.; Elwood, R.W. Trade-Offs between Avoidance of Noxious Electric Shock and Avoidance of Bright Light in Shore Crabs Are Consistent with Predictions of Pain. *Animals* **2024**, *14*, 770. [CrossRef]
40. Okada, S.; Hirano, N.; Abe, T.; Nagayama, T. Aversive operant conditioning alters the phototactic orientation of the marbled crayfish. *J. Exp. Biol.* **2021**, *224 Pt 6*, jeb242180. [CrossRef] [PubMed]

41. Appel, M.; Elwood, R.W. Motivational trade-offs and potential pain experience in hermit crabs. *Appl. Anim. Behav. Sci.* **2009**, *119*, 120–124. [CrossRef]
42. Elwood, R.W.; Appel, M. Pain experience in hermit crabs? *Anim. Behav.* **2009**, *77*, 1243–1246. [CrossRef]
43. Magee, B.; Elwood, R.W. Trade-offs between predator avoidance and electric shock avoidance in hermit crabs demonstrate a non-reflexive response to noxious stimuli consistent with prediction of pain. *Behav. Process.* **2016**, *130*, 31–35. [CrossRef] [PubMed]
44. Fossat, P.; Bacqué-Cazenave, J.; De Deurwaerdère, P.; Delbecque, J.-P.; Cattaert, D. Anxiety-like behavior in crayfish is controlled by serotonin. *Science* **2014**, *344*, 1293–1297. [CrossRef] [PubMed]
45. Fossat, P.; Bacqué-Cazenave, J.; De Deurwaerdere, P.; Cattaert, D.; Delbecque, J.-P. Serotonin, but not dopamine, controls the stress response and anxiety-like behavior in the crayfish *Procambarus clarkii*. *J. Exp. Biol.* **2015**, *218*, 2745–2752. [CrossRef] [PubMed]
46. Elwood, R.W.; Adams, L. Electric shock causes physiological stress responses in shore crabs, consistent with prediction of pain. *Biol. Lett.* **2015**, *11*, 20150800. [CrossRef] [PubMed]
47. Appel, M.; Elwood, R.W. Gender differences, responsiveness and memory of a potentially painful event in hermit crabs. *Anim. Behav.* **2009**, *78*, 1373–1379. [CrossRef]
48. Britsih Veterinary Association. BVA Policy Position on the Recognition of Animals as Sentient Beings. Available online: https://www.bva.co.uk/media/4052/bva-policy-position-on-the-recognition-of-animals-as-sentient-beings.pdf (accessed on 9 May 2024).
49. Animal Welfare (Sentiece) Act 2022. Available online: https://www.legislation.gov.uk/ukpga/2022/22/enacted (accessed on 3 March 2024).
50. Comstock, G. Pain in Pleocyemata, but not in Dendrobranchiata? *Anim. Sentience* **2022**, *7*, 13. [CrossRef]

animals

MDPI

Review

Behavioural Indicators of Pain and Suffering in Arthropods and Might Pain Bite Back?

Robert W Elwood

School of Biological Sciences, Queen's University Belfast, Belfast BT9 5DL, UK; r.elwood@qub.ac.uk

Simple Summary: Pain is an unpleasant emotional state that produces behavioural changes to minimize future tissue damage and promote recovery and survival. These behavioural changes have been demonstrated in crustaceans, insects, and, to a lesser extent, spiders. Other arthropod groups have received little attention with respect to pain. The examination of situations in which individuals might attempt to cause pain in order to manipulate others might offer new opportunities for research into pain in arthropods. For example, defensive venom, traumatic mating, and fighting might inflict pain. This might benefit the animal causing the pain and result in a cost to the animal in pain.

Abstract: Pain in response to tissue damage functions to change behaviour so that further damage is minimised whereas healing and survival are promoted. This paper focuses on the behavioural criteria that match the function to ask if pain is likely in the main taxa of arthropods. There is evidence consistent with the idea of pain in crustaceans, insects and, to a lesser extent, spiders. There is little evidence of pain in millipedes, centipedes, scorpions, and horseshoe crabs but there have been few investigations of these groups. Alternative approaches in the study of pain are explored and it is suggested that studies on traumatic mating, agonistic interactions, and defensive venoms might provide clues about pain. The evolution of high cognitive ability, sensory systems, and flexible decision-making is discussed as well as how these might influence the evolution of pain-like states.

Keywords: Mandibulata; Chelicerata; nociception; pain; traumatic mating; contests; venom

1. Introduction

Living arthropods comprise two major groups, the Mandibulata (including crustaceans, insects, centipedes, and millipedes) and the Chelicerata (including spiders, scorpions, and horseshoe crabs) [1]. They are named after their biting appendages, which differ in origin. The mandibulate homolog of the chelicerae is a pair of antennae. By contrast, the chelicerate homolog of the mandibles is the first pair of walking legs.

Arthropods have a hard exoskeleton that surrounds the muscles and other soft tissue. They are typically segmented and possess jointed appendages that vary in function. The phylum arose in the Cambrian explosion circa 530–550 million years ago and evolved numerous clades, many of which died out (e.g., trilobites). The survivors, however, now comprise the most speciose animal phylum. They are found in marine, freshwater, and terrestrial habitats and are often highly mobile, including flight in some insects. Many arthropods have good visual abilities, albeit with different types of eyes. Chelicerates typically have eyes comprising a single lens and associated "retina" of light-sensitive rhabdomeres, but they often have numerous eyes. By contrast, the mandibulates typically have fewer eyes but each eye may be compound, with hundreds or thousands of ommatidia. Each ommatidium contains a lens and a cone that funnels light to a photosensitive organ. The long, thin ommatidia are bunched together, with each ommatidium pointing in a slightly different direction, which allows for good acuity, and they can detect the polarization of light. Arthropods have a plethora of other sensory modalities. For example, there is the widespread use of olfaction, exemplified by the remarkable perceptual abilities of male moths to detect potential mates over long distances via female pheromones [2]. These modalities enable arthropods to gather various sources of information to enable effective decisions that have important fitness-related consequences, such as locating food, nest sites, mates, dealing with competitors, navigation, and avoiding predators.

Arthropods also detect noxious stimuli, such as heat, mechanical pressure, and chemicals [3–5]. These stimuli are detected by nociceptors that have naked nerve endings within the integument. Nociceptors are key to protecting animals from tissue damage and are found in many phyla, including annelids, molluscs, and chordates, as well as arthropods. The firing of nociceptors might initiate a nociceptive reflex that causes the swift withdrawal of the animal from the locality of the noxious stimulus. Alternatively, just the afflicted part of the body, for example, a leg or antenna, may be moved. Such observations demonstrate that the animal can perceive those stimuli and that the stimuli are aversive.

In humans, exposure to noxious stimuli might lead to a negative affective state that we call pain. Pain is defined by the International Association for the Study of Pain as: "An unpleasant sensory and emotional experience associated with, or resembling that associated with, actual or potential tissue damage" [6]. Pain is presumed to be widespread amongst the chordates and there is evidence consistent with pain in cephalopod molluscs [7] and some arthropods [5,8,9]. However, the idea of pain in non-mammalian groups is dismissed by some, for example, Diggles [10] for crustaceans, or regarded as highly unlikely, for example, Adamo [11] for insects. A frequent argument against pain in these taxonomic groups is that their responses to noxious stimuli are merely nociceptive reflexes. A reflex would not require the involvement of the central nervous system, and without that, there would be no negative affective state. If this is true, then there would be little or no welfare concerns about suffering in arthropods. Certainly, the initial response to tissue damage might involve a reflex but that does not negate the possibility of a subsequent pain experience, as seems to occur in humans [8]. This makes assessing pain difficult because we cannot rely on immediate withdrawal reactions to suggest pain. Further, pain is a private experience, and arthropods cannot report their feelings. Thus, criteria are required that should be fulfilled for each taxonomic group before we might suggest if pain is possible; this is the broad approach taken here.

2. Possible Criteria for Pain

Various sets of criteria have been proposed, some of which were aimed primarily to examine vertebrates [12,13], some were created primarily with invertebrates in mind [8,9,14], and others attempted to cover all animals [15]. These sets of criteria differ with respect to which criteria are used, and there is limited agreement about which criteria are useful [9,16–18]. Here, it is proposed to start by asking questions about the expected function of pain and ask if this might provide guidance about which are the most suitable criteria.

Broadly speaking, the function of pain is to alter the long-term behaviour to mitigate the likely reduction in fitness caused by a wound. Thus, pain might cause the animal to promote healing and recovery by guarding or attending to the injury. Pain should enhance the ability of the animal to avoid or lessen subsequent exposure to the noxious, tissue-damaging stimuli. This might involve avoidance learning and a reduction in the willingness to take risks [15]. Anxiety, including heightened responsiveness to some stimuli coupled with risk aversion after tissue damage, might be highly beneficial, especially because wounded animals may be selectively targeted by predators [19,20]. Altered motivation regarding access to or use of resources is expected. Further, pain might cause the animal to give up important resources, but the costs of giving up a resource should be balanced against the benefit of avoiding the pain [21,22]. That is, we expect pain to have marked effects on behaviour in ways that the animal can best cope with the tissue damage and promote subsequent survival [8,12,23]. The six criteria listed below reflect these behavioural attempts to mitigate fitness loss as applied to arthropods:

(1) Swift avoidance learning (including forming preferences for analgesics or local anaesthetics);
(2) Anxiety and risk aversion;
(3) Long-term changes in behaviour not easily ascribed to associative learning;
(4) Trade-offs between avoidance of the noxious stimulus and other motivational requirements;
(5) Activities directed specifically towards the site of damage (rubbing) and reduction in the use of specific appendages (as in limping);
(6) Protection from further damage by limb autotomy.

These criteria include behavioural changes that go beyond those of nociceptive reflexes. However, other criteria have been proposed that relate to mechanisms by which these behavioural changes are achieved [15,19]. They include physiological changes, identifying brain structures that mediate behaviour change, the presence of nociceptors, and the descending control of nociceptive input. Whilst these are interesting, they are not as useful as criteria for pain as the behavioural changes they support. Some of these secondary criteria may be deduced from the primary, behavioural observations. For example, one suggested criterion is that the animal should possess brain mechanisms that allow nociceptive input to be integrated with other sensory inputs [9]. However, if an animal changes its behavioural response to the noxious stimulus depending on other motivational needs, i.e., a trade-off, then it must have the mechanism to do that [16,17]. When discussing the possibility of pain in arthropod taxa these secondary criteria will be mentioned but those relating to behaviour will be given prominence.

The approach used here will not employ a formal scoring system like that of Crump et al. [9] and Gibbons et al. [24]. That is because there is little agreement about the weighting of scores from different criteria [17,18] or about the inclusion/exclusion of criteria [16]. Rather, a more relaxed approach will be used that notes whether there is evidence that supports each behavioural criterion. Experiments designed to test if behavioural criteria are upheld will be reviewed, as well as anecdotal evidence for and against pain in particular taxa. Other observations and phenomena will then be considered that might help to understand if pain occurs in arthropods. First, cases will be examined in which the male inflicts tissue damage to the female during mating and if that dissuades the female from mating with another male. Second, it is asked if pain might be inflicted during fights for specific resources. Third, venoms used for defence are examined and it is asked if they act against arthropods and if the defence might involve inducing pain. The aim will be to encourage broad investigations of pain that use different sources of information and suggest new approaches for pain research. Finally, because high cognitive ability has been suggested to be required for a pain experience, examples within the arthropods will be briefly reviewed. This will be viewed in an evolutionary framework to help assess which arthropod groups might benefit from the emotional experience of pain [25]. The possible role of consciousness in pain experience is beyond the scope of the present paper and good references on this are available [9].

3. Evidence

3.1. *Mandibulata*

3.1.1. Crustaceans

Crustaceans show swift avoidance learning when repeatedly exposed to noxious stimuli. For example, marbled crayfish (*Procambarus virginalis*) in a T-maze initially showed a preference for an arm with blue light compared to an arm with white light. However, if the former resulted in electric shock, the preference was switched, in the short-term, after just one exposure to shock. Further, three exposures resulted in avoidance that lasted for 48 h [26]. Long-term avoidance was not observed, however, if the animal was subsequently kept at low temperatures or was given protein-formation inhibitors, suggesting that long-term memory was inhibited. Shore crabs (*Carcinus maenas*) offered a choice of two dark shelters in which to avoid bright light quickly selected one [27]. If that shelter was accompanied by electric shock, it failed to result in avoidance of that shelter in the next choice trial; rather, the crabs showed a strong preference to return to their first choice. However, in the following choice trial, those crabs that received a shock in the previous trial switched their choice to the alternative, safe shelter, thus indicating swift avoidance learning. Tests in later trials demonstrated that the crabs discriminated shelters by the direction of walking from the start point (left or right) rather than visual cues. Crabs that were turned around by 180° in subsequent tests walked to the wrong shelter [27]. However, when a different paradigm was used that involved crabs only having access to one shelter at a time over ten trials, they failed to avoid the shelter that was previously accompanied by electric shock in a final choice test when both could be accessed [28]. This is in keeping with other studies that showed that the sequential presentation of stimuli is less effective than simultaneous presentation in learning studies [29]. Nevertheless, the crabs appeared to show other forms of learning that reduced their exposure to shock because in later trials they escaped from the shock shelter significantly earlier than they did in early trials [28]. Other studies that demonstrate avoidance learning are reviewed by Crump et al. [9]. However, the author is not aware of studies on crustaceans that have tested possible preferences for analgesics or local anaesthetics when exposed to noxious stimuli.

Increased anxiety (wariness) following noxious stimuli has been demonstrated in crayfish, *Procambarus clarkii* [30]. These animals show a preference for dark environments, and when placed in a cross maze with two light and two dark arms, they spent more time in the dark than in the light arms. However, when they were given repeated electric shocks that induced escape responses, prior to being placed in the cross maze, they showed an enhanced avoidance of the light arms. This 'anxiety' was accompanied by higher levels of serotonin (5HT) in the brain [30,31]. Further, animals that were injected with 5HT, rather than being shocked, showed similar levels of anxiety to those that were shocked. Surprisingly, chlordiazepoxide (CDZ), a drug that reduces anxiety in humans, reduced signs of anxiety in crayfish. Aversive visual stimuli also induced anxiety in the crab, *Neohelice granulata*, and that state was blocked by a serotonin inhibitor [32]. Behaviour suggesting anxiety after receiving shocks was also noted in the amphipod *Gammarus fossarum* because the amphipods increased their time hiding in shelters and decreased their time swimming freely [33]. The fitness benefit of this anxiety was shown by increased survival in the presence of predatory fish [33]. The anxiolytic LY354740 reduced the time spent hiding in a dark shelter. It is interesting to note that all three studies showed that drugs used to treat anxiety in humans were effective in crustaceans, suggesting similar anxiety mechanisms in various taxa.

Long-term behavioural change caused by electric shocks on the abdomen, within the shell, has been examined in hermit crabs (*Pagurus bernhardus*) [22]. Shocked crabs and non-shocked controls were later offered a new empty shell. Those that had been shocked approached the empty shell in a shorter time and engaged in a shorter investigation prior to moving into that shell compared with non-shocked crabs. This indicated that shock within the shell caused the crab to value that shell less than did crabs that were not shocked [34,35], and were thus highly motivated to seek and take a new shell. In one experiment, the new shell was offered 20 s after the cessation of shock (or an equivalent time for the non-shocked group). A second experiment, however, used longer times between shock and the offering of a new shell [36]. This provided evidence of motivational change that lasted at least 24 h because the crabs still showed an enhanced motivation to change shells (Figure 1).

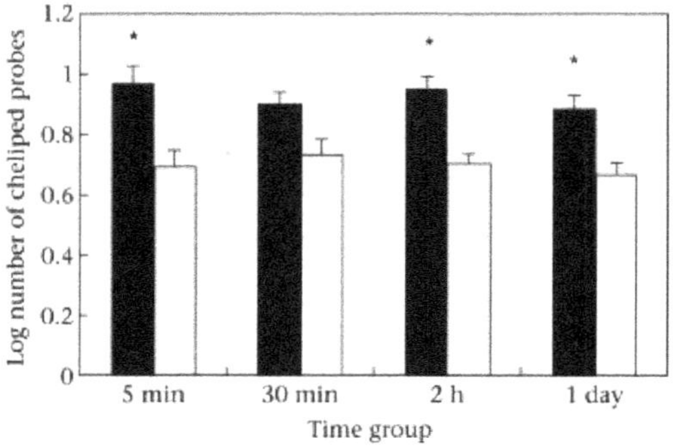

Figure 1. Hermit crabs that were shocked within their shells (light bars) investigated new empty shells using fewer cheliped probes before occupying those shells than did non-shocked crabs (dark bars) ($F_{1,112}$ = 33.528, $p < 0.0001$). The times between the shock and the new shell being offered were varied and the effect of shock was significant (asterisks denote $p < 0.05$) for the 5 min, 2 h, and 1-day groups. From Appel and Elwood 2009 [36].

Swift escape from noxious stimuli might appear to be a nociceptive reflex, however, it might be more complex than that [37]. This possibility was tested by assessing if trade-offs occurred between escape responses and other motivational requirements [21,22]. The rationale was that if these immediate responses to a noxious stimulus were influenced by other factors, then the animal must have integrated information from other sources with that from nociceptors, resulting in an adaptive decision rather than an inflexible reflex [9,16]. For example, some hermit crabs (*P. bernhardus*) left their gastropod shells if shocked within their shells, however, they evacuated less preferred species of shell at a lower voltage than a preferred species [21] and, if the voltage was kept constant, they were more likely to evacuate the less preferred species [22]. That is, the benefit of escape was balanced against the cost of giving up a specific resource. Further, hermit crabs were less likely to leave their shell if the odour of a predator was in the surrounding water [38]. In this case, the balance was between escaping from the noxious stimulus and the risk of predation.

A further example of decapods modifying their escape response comes from video recordings of crayfish (*Procambarus clarkii*) when touched with a hot probe [4]. Rather than all crayfish showing a similar reflex withdrawal, some grabbed the shaft of the probe in a defensive action. That is, the response was clearly influenced by other factors and might reflect a trade-off between defence and escape [37]. These studies demonstrate that even short-term or immediate responses may not be reflexes but rather are the result of centrally organized decisions that maximize fitness (on average) following tissue damage.

Rubbing, guarding wounds, and limping are interpreted as being consistent with pain [8,9,37]. For example, sodium hydroxide or acetic acid (both known to induce pain in mammals) applied to a single antenna of the glass prawn (*Palaemon elegans*) resulted in prolonged grooming and rubbing [3]. Grooming involved repeatedly pulling the antenna through the small chelipeds (claws) or the mouthparts, whereas rubbing was pressing and moving that antenna against the side of the tank. The responses were directed at the treated antenna significantly more than the untreated antenna, indicating an "awareness", or at least the ability to locate the site of the noxious stimulus [37]. Further, the application of sodium hydroxide to one eye of a glass prawn caused high levels of grooming of that specific eye with either one or both first walking legs. This behaviour was not seen if just seawater was applied [8]. Similar directed activities occurred in shore crabs (*C. maenas)* when they used their claws to scratch at their mouthparts that had been brushed with acetic acid [39]. Hermit crabs that got out of their shells after being shocked on their abdomen often groomed at that site, a behaviour not seen if the crabs were removed from the shell by other means [36]. Further, edible crabs (*Cancer pagurus*) that had a clawed appendage twisted off, in the manner used in some fisheries [40], held their remaining claw over the wound during competitive interactions. Finally, when formalin was injected into a claw of a shore crab (*Hemigrapsus sanguineus*), the crab reduced the use of that appendage when walking and often pressed that claw against its carapace [41]. They also shook and rubbed the injected claw [41].

One way of mitigating tissue damage to appendages in crustaceans is to autotomise that limb. This is achieved by muscular contraction so that the limb is cast off on a fracture plane, which quickly heals. Cutting a distal joint of a claw of edible crabs (*C. pagurus*), which caused haemolymph loss, reliably induced autotomy [42]. Autotomy was also seen after claws of shore crabs (*H. sanguineus)* were injected with formalin [41] and autotomy also occurred when the walking legs of *C. maenas* were injected with acetic acid [8]. The observation of autotomy after tissue damage and loss of haemolymph might be attributed specifically to that fluid loss, however, the injection of substances known to cause pain in mammals does not result in fluid loss, yet autotomy still occurred. Further, crabs autotomized limbs when the whole animal was placed on a hot plate [43] or if the leg was subjected to electric shock [44]. These observations are consistent with the idea of pain mediating the autotomy response [8]. Further research is required, however, to determine if autotomy is a behavioural decision or simply a reflex response. That is, we need to determine if autotomy can be modified by motivational factors other than the noxious stimulus.

3.1.2. Insects

Numerous studies demonstrated avoidance learning in insects and these are reviewed in detail by Gibbons et al. [24] and Pitman et al. [45]. For example, fruit flies (*Drosophila melanogaster*) avoided an odour that had been associated with electric shocks, even after just one trial, and they showed retention of the learning after 24 h [46]. Honeybees (*Apis mellifera*) associated electric shock or heat with preceding cues and similar classical conditioning has been shown in several other insects [24]. Operant conditioning to avoid noxious stimuli has also been shown, for example, in cockroaches (*Periplaneta americana*) and honeybees that learned to avoid locations in which they received electric shocks [47,48]. Other forms of learning such as trace conditioning and reversal learning probably require more complex cognitive abilities than classical or operant conditioning. In trace conditioning, there is a temporal gap between the conditioning stimulus and the noxious event, which implies a memory of the CS, and this has been shown in fruit flies [49]. Reversal learning requires the subject to learn about the switching of the predictive value of the stimuli. This could be achieved by the animal starting again in learning the task after each switch or it could learn that things can switch and thus speed up the switch in behaviour, this being demonstrated in different insects [24].

There has been one study on the preference for analgesics by bees (*Apis mellifera*) [50]. Injured and uninjured bees were offered a choice between a sucrose solution and sucrose with morphine, however, there was no difference in the preference for morphine even though injured bees took more food overall. Thus, there is no evidence for the self-administration of analgesics. One problem with the experiment, however, is that there is no evidence of morphine receptors in insects and thus no evidence that morphine has an analgesic effect [51].

One way to examine anxiety or risk aversion is to use cognitive bias tests [52]. Here, a subject receives a positive reward from one stimulus and no reward or a negative outcome with another stimulus. This may involve mixtures of two odours with a mix of 1:9 or 9:1 as the two stimuli. When this was conducted with honeybees, the bees learned which mix produced the more positive outcome. Then, some bees were shaken to induce anxiety, and their responses were compared to those of undisturbed bees. The bees were then offered ambiguous stimuli (mixtures of 3:7, 1:1, and 7:3) to determine if the bees approached these. It was shown that shaken (anxious) bees were less likely than the unshaken bees to approach these ambiguous stimuli, suggesting emotional states in these insects [52]. Similarly, bees subjected to simulated attacks by cryptic predators became more wary when foraging and lost feeding opportunities due to false alarms [53]. However, these studies did not specifically link these findings to possible pain.

Long-term behavioural change was examined in *Drosophila* that had a middle leg amputated. They subsequently showed a greater number of escape responses when placed on a surface at 38 °C than did intact animals, but only between 5 and 21 days after injury [54]. This was described as consistent with thermal allodynia, where a "painful" behavioural response is elicited by normally innocuous stimuli.

Trade-offs between avoidance of noxious heat and the requirement for quality food were examined in bumblebees (*Bombus terrestris*) [55]. Bees were allowed to forage in an arena where there was a choice between two high-quality feeders (containing a 40% sucrose solution) and two alternative feeders. The types of feeders also had colour cues. Four experimental groups of bees were each offered either a 10%, 20%, 30%, or 40% sucrose solution in the alternative feeders. The bees were allowed to forage when the feeders were all at room temperature and the bees learned to use the higher concentration feeders. Subsequently, the high-quality feeders were heated to a noxious level, but how much these were avoided depended on what was in the alternative feeders. If those feeders had low concentrations of sucrose, the avoidance of the heated feeders was lower than if the alternatives had high concentrations. That is, the bees traded off food quality with avoidance of a thermal noxious stimulus.

Directed grooming at the site of a noxious stimulus has also been reported for bumblebees (*B. terrestris*) [56]. Bees were either touched on one antenna with a heated iron or with a non-heated iron and then observed to determine which antenna (touched or untouched) and type of touching (heated or nonheated) received the most grooming. In the two minutes following the touch, there was little antennal grooming of the untouched antennae, and little directed towards the touched antenna if the iron was not heated. However, there was significantly more grooming of the specific antenna touched with the heated iron. This indicates that touching per se is not noxious but touching with a heated iron is, and the bees directed their grooming towards the site of the noxious stimulus.

Autotomy occurs in many insects, for example, field crickets (*Gryllus bimaculatus*), various leaf-footed bugs (Hemiptera: Coreidae + Alydidae) [57], and stick insects (*Didymuria violescens*) [58]. It occurs when a leg is damaged due to interspecific competition or failed predation attempts or when the leg is held [57]. Autotomy following damage is consistent with the expectation of pain.

The idea that insects might experience pain is frequently dismissed because of observations of apparently normal behaviour continuing while the insect is being damaged. For example, male praying mantids may be eaten by their females during mating but show no attempt to escape [59]. It is argued that if the male could feel pain, then it would not continue with copulation. However, we should consider the fitness benefits to the male if he escaped and the possible loss of fertilizations from that attempted mating [60]. The male might survive but if it cannot mate successfully with another female to offset the lost fertilizations with the current female, then it might lose the chance of siring a brood. Thus, it might be beneficial in genetic terms not to struggle so that they ensure copulation continues with the successful transfer of sperm. However, males that have yet to initiate copulation with the female may struggle to escape the predatory attempts of females [60]. In this situation, the males show vigorous waving and pushing with the legs in apparent attempts to escape, which is consistent with nociception and possibly pain. The observation of some males not struggling during mating when being bitten, but others showing vigorous escape responses when attacked outside of mating, suggests that males may suppress the neural control of the response to tissue damage specifically when mating [15,51]. This modulation of responses suggests a trade-off between escape and the opportunity to mate; it deserves further investigation, as does the possibility of the descending control of nociception/pain [51].

3.1.3. Centipedes and Millipedes

Centipedes are fast-moving predators whereas millipedes are slow-moving and consume mostly vegetable matter. No studies have been made that might suggest pain in these animals other than the autotomy of legs being noted in centipedes [61]. Thus, there is ample scope for research on these two groups.

3.2. Chelicerata

3.2.1. Spiders

Several experiments are directly relevant to investigating the possibility of pain in spiders. For example, avoidance learning has been demonstrated in the wolf spider (*Schizocosa avida*) when they suffer damage when escaping from a predatory attempt by a scorpion; they subsequently avoided the odours of such scorpions [62]. Jumping spiders (*Hasarius adansoni*) also learned to avoid visual stimuli associated with high temperature [63] and electric shock [64,65]. Thus, spiders appear to learn from noxious experiences and then reduce or avoid those noxious events in the future [66]

Leg autotomy also suggests pain in spiders [8]. For example, in *Argiope aurantia* [67], autotomy was noted when these spiders attempted to capture ambush bugs (*Phymata fasciata*), but the bug grasped a spider leg and probed a joint with its proboscis. Experimental penetration of the joint with a sterile pin did not cause autotomy, indicating that the saliva of the ambush bug likely had an effect. The venomous saliva of the bug is painful to humans, suggesting pain may play a part in autotomy [67]. When bee and wasp venom were injected into a spider leg, they induced autotomy [67]. Individual components of bee venom were then injected, some of which caused autotomy. The effective components were histamine, serotonin, phospholipase, and melittin, all of which induce pain in humans. The ineffective components were acetylcholine, bradykinin, hyaluridase, adrenaline, and dopamine. Acetylcholine and bradykinin induce pain in humans but not autotomy in spiders, and hyaluridase, adrenaline, and dopamine do not. Thus, substances that induce pain in humans are likely to induce autotomy in spiders, whereas those that do not cause pain in humans do not cause autotomy.

3.2.2. Scorpions

Reports of experiments and observations that are consistent with the idea of pain in scorpions are not common. However, avoidance learning has been noted. When giant whip scorpions, *Mastigoproctus giganteus*, were tested in a shuttle box in which electric shock could be applied to one side at a time, they learned to shuttle between compartments to avoid the electric shock [68]. This indicates the aversive nature of the shock for these animals and that they swiftly learn how to escape.

3.2.3. Horseshoe Crabs

Atlantic horseshoe crabs (*Limulus polyphemus*) are used in the biomedical industry to check that injectable vaccines and medicines are safe from contamination by endotoxins. The blood from these animals is collected from wild-caught animals and the animals are subsequently released back to the wild. There is concern about the ethical aspects of this practice, but much relates to the level of mortality following bleeding and release and, hence, the sustainability of the populations [69]. Whilst there is concern about the potential for adverse welfare effects on individuals, very little is known about the potential for pain. Bleeding results in a considerable reduction in the remaining hemocyanin levels and mortality increases and activity levels decline with reduced blood [70,71]. Attempts to examine avoidance learning are limited and there is no convincing behavioural evidence that might be consistent with the idea of pain [72]. Given the high use of these animals, it is surprising that so little empirical evidence about sentience is available.

4. Discussion

This review shows that behavioural observations consistent with pain are not equally distributed among the various arthropod groups (Table 1). They are clustered primarily in the crustaceans and insects of Mandibulata, and to a much lesser extent in spiders of Chelicerata. They are virtually absent in millipedes and centipedes of Mandibulata and the scorpions and horseshoe crabs of Chelicerata. Hopefully, these neglected groups will soon receive attention to close these gaps in knowledge.

Table 1. The behavioural criteria for which there is evidence is shown for each taxon.

Behaviour	Mandibulata				Chelicerata		
	Crustacean	Insect	Centipede	Millipede	Spider	Scorpion	Horseshoe Crab
Avoidance	√	√			√	√	
Anxiety	√	√					
Long-term changes	√	√					
Trade-offs	√	√					
Directed activities	√	√					
Autotomy	√	√	√		√		

For crustaceans and insects, there is evidence consistent with each of the six behavioural criteria. There are examples of avoidance learning, anxiety, and risk aversion, long-term changes in behaviour not easily ascribed to associative learning, trade-offs between avoidance of the noxious stimulus and other motivational requirements, activities directed specifically towards the site of damage (rubbing) and reduction in the use of specific appendages (as in limping), and protection from further damage by limb autotomy. However, there is no evidence that suggests preferences for analgesics or local anaesthetics. For spiders, there are examples of avoidance learning and autotomy. For centipedes, there is evidence for autotomy, and for scorpions, there is support for avoidance learning. For millipedes and horseshoe crabs, there are no studies that support any criterion.

It should be acknowledged that finding evidence consistent with the idea of pain is not the same as proving pain [8,37]. Nevertheless, the accumulation of evidence for sentience in decapod crustaceans, as outlined by Birch et al. [14], was sufficiently compelling for legal changes to be made in the UK which now recognise the potential for pain and suffering in decapods (Animal Welfare (Sentience) Act 2022). Further, using the set of criteria of Birch et al. [14], Gibbons et al. [24] suggested that there is strong evidence of sentience in several orders of insects. However, those reviews used a mix of behavioural and neurological criteria to support their cases. They used some of the behavioural criteria used here, but they also examined the evidence regarding the presence of nociceptors, of brain regions that might integrate different sensory inputs and regions that integrate nociception and other sensory modalities and analgesic effects. As noted previously, the first three of these are demonstrations of the mechanisms that enable functionally important behavioural changes.

Nociceptors are found in most metazoans but have been particularly well-described in insects, especially *Drosophila* [24]. Nociceptors have also been shown to occur in crustaceans [9,14]. The brains of arthropods show significant functional localization, and areas involved in the integration of different sensory modalities, including nociception, have been described [9,24,73]. For example, decapods have well-defined circuits and centres that allow for the integration of nociception with other sensory inputs and between different sensory modalities [74] and for links with learning centres [75]. These aspects are excellently reviewed by Crump et al. [9] who concluded that decapods have the neural capacity that would be expected if they experience pain. There is also extensive work on insect brains, with reference to the structure and abilities that integrate different sensory systems and integrate nociceptive input with other sensory inputs [24]. Further, the effects of local anaesthetics have been noted in crustacea [3,14] and insects [24]; however, the analgesic effects of opioids in these two groups remain in doubt [24,44]. Studies of anaesthetics and analgesics help shed light on the physiological mechanisms underpinning nociception in arthropods. All these observations are important in understanding the biology of this phylum. However, these previously suggested criteria for pain are related to the mechanisms mediating the behavioural changes that follow tissue damage, but they cannot be regarded as filling the function of pain.

Physiological stress responses have been proposed as a criterion of pain [8,15,16] but were not used as such by Birch et al. [14]. For example, stressed crustaceans show elevated crustacean hyperglycaemic hormone (CHH) [76], whereas stressed insects show changes in biogenic amines (octopamine and dopamine), neuropeptides (allatostatin and corazonin), and metabolic hormones (adipokinetic and diuretic hormones) [77]. These systems have similarities to the cortisol of vertebrates. In decapods, CHH causes the release of glucose from stores of glycogen [78] and increases levels of lactate. Breaking a claw of edible crabs (*C. pagurus*) results in a swift increase in glucose and lactate in the haemolymph [42]. Pain is likely to cause stress so measuring physiological changes might provide a way of assessing pain.

However, there is a problem with using physiological changes to investigate pain because noxious stimuli typically result in escape responses and other vigorous activities that might cause physiological change. For example, in the study of anxiety in crayfish, the electric shocks used to stress the animals caused repeated tail-flick responses [30,31]. However, one study [79] attempted to distinguish between the direct stress of electric shock in causing elevated lactate from that caused by increased activity. Some shore crabs (*C. maenas*) received electric shock via wires attached to walking legs and others were wired in the same way but without shock. Some of those receiving shock showed increased activity such as threat responses, but most walked around the enclosure. Some of those not receiving shock remained stationary during the test but most walked around the enclosure. When lactate levels were examined in those animals that just walked about the enclosure, those that received the shock were substantially higher than those that were not shocked. Thus, the stress response was specifically due to the shock rather than the behavioural responses. This is consistent with the idea that the shock induces a pain-like state that is stressful to the crabs [79]. Although behavioural responses have been the focus of this review, physiological changes are also specific responses to noxious stimuli and modulate further behavioural responses to injury [20] and, thus, might be a reasonable criterion for pain.

5. Do Animals Inflict Pain on Others to Gain an Advantage?

Should a pain system evolve, then individuals might be susceptible to being manipulated by being subjected to pain by other animals. Here, three situations are considered that might occur in arthropods.

5.1. Tissue Damage Caused during Mating

There are many examples of male copulatory traits that reduce the chances of the female mating with a subsequent male [80]. These include copulatory plugs that impede a subsequent male and the production of chemicals that reduce the motivation of a female to mate or perhaps reduce her attractiveness to other males. Another mechanism involves the male causing tissue damage to the female to reduce her willingness to mate again [81]. One example is the evolution of spines on the male genitalia of bean weevils (*Callosobruchus maculatus*) that cause tissue damage to the female [82]. Females have reduced longevity due to the damage and act to minimise that damage by kicking at the male during mating attempts. The tissue damage might thus act as a "disincentive for unfaithfulness" [80].

That is, the tissue damage might change the long-term motivation of the female in a way that would be consistent with the idea of pain.

A second example relates to the mating of a wolf spider (*Pardosa pseudoannulata*) [83]. During insemination, the inner walls of the female genital tract are damaged with the sharp intromittent organ of the male. Females then show a reduced willingness to mate with a subsequent male. This might be due to chemical manipulation [80], but experiments demonstrated that the seminal fluid had only minor effects. Experimental damage by microinjection without seminal fluid, however, greatly suppressed mating [83]. This implies that the female does not require memory of mating or a male to show a reduction in the willingness to mate because with microinjection neither is involved. It does suggest, however, that tissue damage per se reduces the willingness of the female to mate. The long-term shift in motivation is a key expectation of pain.

A third example is found in scorpions. Male scorpions use their chelae to take hold of the female's pedipalps and attempt to position her so she can take up a spermatophore he has placed on the substrate [84]. During this process the male may sting or club the female, possibly to subdue a reluctant female. Some species show sexual dimorphism in the stinger and the components of the venom, suggesting that these might be important for male success in mating [85]. One might speculate that males might also benefit by inflicting painful stings that reduce subsequent mating by that female.

Sometimes the conflict between the sexes might result in female countermeasures as is seen in female damselflies, *Enallagma cyathigerum*, that have a conspicuous vulvar spine which contacts with the male during copulation [86]. Males copulated for longer with non-virgin females that had the spine removed, showing that the spine reduced the duration of mating and might reduce the male's ability to remove the sperm from a previous copulation. It was proposed that the "spine allows females to exert some control over copulation duration by producing enough "discomfort" or "pain" to males to reduce copulation duration" [86].

5.2. Possible Pain Inflicted to Win Fights over Resources

It has been suggested that intraspecific contests might alter the emotional state of the participants [87] and that losers show a negative affective state, which is presumably induced by the agonistic actions of the opponent. One interpretation of this effect is that the fight activities might induce pain in one or both opponents. For example, mantis shrimps, *Neogonodactylus bredini,* use their predatory appendages to strike each other during contests, often involving attempts by one to block the blows of the other with their armoured telson. Although obvious wounding was not observed, these blows are known to influence the motivation of an opponent to stay in the contest [88]. Another example of striking is seen in the shell fights of hermit crabs, which hit their shell against that of an opponent. The power of this shell rapping has a direct effect on the motivation of the opponent, and, thus, the probability of it giving up and abandoning the shell [89]. It shows similarities to the observations of hermit crabs receiving electric shocks on the abdomen, which can induce shell evacuation, consistent with ideas about pain [21,36,90]. However, decisions in fights are not just about the ability of opponents because losers typically persist for longer when they are contesting a high-value resource [91]. Arthropods (and other animals) accept higher costs in fights over high-value resources and possibly fight to the death when the resource is vital for reproduction [91]. That is, with high-value resources, we expect to see descending control of nociception (and pain) during fights, but we expect to see less descending control with lower-value resources. This would match observations about the trade-offs between pain avoidance and keeping valued resources that have been described above for crustaceans and insects. Thus, pain would provide a mechanism by which fight decisions are optimised, not only in arthropods but in a wide range of other animals.

5.3. Chemical Warfare and Possible Pain

Many arthropods may cause pain to mammals and this ability might shed light on the issue of arthropods feeling pain. Some use stings (e.g., scorpions, wasps, bees, and ants), some may use a proboscis (e.g., assassin bugs), and some may bite and inject venom (e.g., centipedes and spiders). Some may spray chemicals, as in bombardier beetles and ants, and some caterpillars have venomous spines. Most of the interest in these arthropods is about their effects on humans, who may suffer pain and possible death However, these arthropods have a long evolutionary history, and the venoms may pre-date humans and most mammals. The venoms have evolved for two functions, that is, to kill prey and deter potential predators, and these are likely to include arthropods. They may also be used in intraspecific contests. Ants also use chemical sprays that deter other species of ants [92].

There is considerable variation in the venom within wasps and bees although there are some common components [93]. The differences in components often relate to the lifestyle of the species. For example, solitary wasps are predatory and use their stings to subdue, paralyze, or kill their prey. However, bees evolved a non-predatory life from wasp ancestors but retained the sting [94]. Here, the sting is often used to deter potential predators, particularly important in social species, and these venoms cause pain in mammals [93,95]. They also use stings, however, in defence against other invertebrates and in intraspecific contests. The question posed here is: Could the evolution of these mechanisms be driven (at least in part) by the benefits of inflicting pain on arthropods? As previously mentioned, the components of bee venom that cause pain in humans may induce autotomy in spiders, whereas those that do not cause human pain do not induce autotomy in spiders [67].

Scorpions use their stings in predation and defence, and they use the venom in a judicious manner [96]. They might use dry stings in defence when the risk is low, and when the risk increases, so does the volume of venom [96]. Further, scorpions may alter the composition of the venom and use components that lead to the death or subjugation of prey but select components that cause pain (at least in mammals) when trying to deter an attack by another animal [97].

Assassin bugs (*Pristhesancus plagipennis* and *Rhynocoris iracundus*) also have two main types of venom, one used in predation and another used in defence [98,99]. The predation venom causes paralysis and helps to break down the tissues of the prey to a liquid form. The defensive venom has components that may trigger nociceptors and might cause pain to deter a potential predator. Thus, the function of these components might be to cause pain to other arthropods. Insertion of the proboscis of assassin bugs causes pain in humans and autotomy in spiders.

Spiders inject venom when they bite and it contains a substantial array of components [100]. Many of these activate nociceptors and result in pain (in humans) but there can also be antinociceptive compounds that have the opposite effect. The functional significance of these substances that influence nociceptors is not clear and it is not known if they are used to deter arthropod predators [100]. Most attention is directed toward their use against vertebrate predators, but it is estimated that the venoms were originally used against arthropods [101].

Bombardier beetles produce a noxious spray that is used to deter potential predators such as ants, spiders, or praying mantids, and the spray might also cause attacking spiders to autotomise legs or mantids to groom the body area hit by the spray [102,103]. The spray of bombardier beetles is the result of mixing two sets of chemicals ordinarily stored separately in the glands. One gland contains hydroquinones and hydrogen peroxide while the other contains catalases and peroxidases [104]. The beetle mixes the contents of the two compartments, causing oxygen to be liberated from hydrogen peroxide and the hydroquinones to be oxidized by the freed oxygen. The resulting liquid is released at about 100 °C and propelled at the attacker. Heat is a key component of the deterrent [104] and is likely to trigger nociceptors and possibly induce pain.

Some limacodid caterpillars possess spines that inject venom, and this protects against invertebrate predators [105]. Prior exposure to the venom also induces avoidance by arthropod predators, suggesting that it induces rapid avoidance learning. Some wasps manage to consume these caterpillars by first chewing off the venom spines, and assassin bugs pierce the caterpillar from the other side of the leaf [105]. The venom is painful to humans and has been linked to protection against vertebrate predators by inducing pain, however, they also protect against common arthropod predators.

It would be interesting to know more about the evolutionary history of these defensive venoms and if they induce signs of pain in other arthropods. It might inform us about the possible evolution of sentience in arthropods and how the ability to suffer from pain might be used to the advantage of other animals. It would also be useful to ask questions about pain in studies of intraspecific contests and traumatic mating.

6. Cognitive Ability and Possible Evolution of Pain in Arthropods

High cognitive ability has been suggested to be a prerequisite for pain [8,12,106–111]. High cognitive ability enables large amounts of sensory information to be integrated, enables effective decisions [112], and provides flexibility in the responses to noxious stimuli that are guided by some expectation of outcome. Recently, the ability to choose between possible responses to noxious stimuli, rather than relying on an inbuilt algorithm, has been suggested to lead to the ability to feel pain [25].

The ability to acquire and manipulate information has been shown in various arthropods. For example, hermit crabs gather information about potential new gastropod shells and integrate that with information about the shell they currently occupy before deciding which is the better of the two [34]. They use several sources of information such as the external and internal shape and size using visual and tactile information, and subsequently gather and integrate further information after moving in [113]. Hermits might also select a shell that would normally be avoided if that specific shell allows them to pass through a small hole to escape from a restrictive area, which suggests a degree of awareness of how to solve a problem [114]. Further, hermit crabs appear to be aware of experimentally induced changes to their shell, such as a plastic plate being attached, which impedes passage through small gaps [115]. The crabs seemed to assess the width of the gap and turn sufficiently to pass through. They quickly adapted to the attached plate and made greater turns to get through.

The integration of information is even more complex when hermit crabs fight over the ownership of shells [90]. They gather information about both the shells and the opponent [116–119]. The attacker also monitors changes in its physiological state, which changes dramatically due to the exertions of the major fight activity of shell rapping [120,121], and shifts its fight behaviour according to those changes. Further, defenders assess the vigour of an attack and make decisions about whether to resist, the former involving mobilisation of glycogen stores [122]. They can also remember previous opponents for up to 4 days after an encounter [123]. Remarkably, a potential attacker might not only assess if it can gain from an exchange of shells but also assess if its shell is suitable for a potential opponent before attacking because the opponent might be more willing to exchange shells if it can gain in shell quality [124]. These crustaceans show an excellent ability to gather, manipulate, and use information from multiple sources, indicating a higher cognitive ability than generally recognised [90].

There are various examples of homing by crustaceans [125]. For example, fiddler crabs (*Uca* sp.) appear to show path integration by which turns are taken on the outward path during foraging, presumably by updating vectors and recalling them during the return [126]. There is also an impressive example of long-distance migration in the spiny lobster, *Panulirus argus,* during which they orientate by use of a magnetic sense [127]. Lobsters displaced by 12–37 km were capable of accurate orientation toward their home location. This necessitated a detection system to provide information about the current location and the home place location, coupled with a directional or compass sense to enable the home path to be determined [128].

Insects also show remarkable cognitive abilities and the best known of these are in Hymenoptera, especially bees [129]. It is possible that these abilities arose due to foraging on flowers and the need for minimising energy expenditure while maximising energy gain. The flowers of different species vary in many ways, including colour, odour, shape, ease of access to nectar, and nectar quality, and discriminating between flower types is key to the success of bees. They can learn to discriminate between two types of flowers and to switch preferences between contexts [129]. Bees can even learn concepts such as whether a visual stimulus is symmetrical or not. That is, after being exposed to several stimuli that differ in symmetry, and with either symmetrical or asymmetrical shapes being rewarding, the bee then discriminated entirely new stimuli based on their symmetry [130,131]. Social learning, in which one animal observes the actions of another and then employs those actions to obtain a reward, has also been noted in bumblebees. One experiment provided disks with food placed under a plexiglass sheet. The food could be accessed by pulling a string attached to the disk but very few bees learned this by themselves. However, those observing experienced bees pulling the string employed that method [132]. These examples, and many others, of high cognitive ability in bees have been reviewed by Chittka [129]. Together, they demonstrate remarkable cognitive abilities that go beyond previous expectations.

High cognitive ability has also been noted in spiders. Jumping spiders, for example, adjust their hunting methods depending on the type of prey [133]. The hunting spider (*Portia labiata*) hunts spitting spiders (*Scytodes pallidus*), which are also predators of spiders and thus dangerous. *Portia* gathers information as to whether the spitting spider is carrying eggs in its mouth and is thus less able to defend itself and modifies its attack accordingly [134]. Further, when hunting in complex environments, *Portia* plans the route to the prey and might take detours to avoid obstructions that initially take it away from the prey item [135]. This suggests an ability to comprehend the complex spatial relationships between itself and the prey and possible routes to a goal [136].

These examples of gathering and integrating sources of information that enable effective decisions demonstrate that various taxa of arthropods have the behavioural flexibility that has been suggested as a requirement for pain to be of use [25]. These observations on cognitive abilities add to the evidence concerning the behavioural criteria for pain in crustaceans, insects, and, to a lesser extent, spiders. This accumulation of evidence thus makes pain at least a possibility in arthropods. Next, it is considered how pain might have evolved in this phylum.

7. Evolution

If pain occurs in crustaceans, insects, and spiders, then the parsimonious explanation for the evolution of pain is that there was one evolutionary step that preceded the split between the Chelicerata and Mandibulata, which occurred in the Cambrian, over 500 million years ago [137]. However, there is less evidence of pain in spiders than in crustaceans and insects, and some might consider this insufficient to accept the possibility of pain in that group. If the spiders are thus disregarded, the parsimonious explanation for the evolution of pain is that it evolved before the crustacean/insect split (approximately 500 million years ago) [137]. However, within these groups, evidence for pain is found primarily in certain groups, such as the decapods in the crustaceans [37] and six orders of insects (Blattodea, Diptera, Hymenoptera, Lepidoptera, Orthoptera, and Coleoptera) [24]. A patchy distribution of pain might occur if pain was lost in some lineages. For example, taxa that evolved from a free-living form to a sedentary lifestyle may have reduced their behavioural choices and, thus, there may be no need for free choice and pain. For example, barnacles, especially those that have adopted a parasitic lifestyle, might have very little opportunity to show flexible behavioural responses to noxious stimuli and thus not benefit from a pain system [25]. However, a lack of flexibility might also be expected in basal groups of arthropods because of the limited ability of primitive sensory systems to gather information [25]. Improvements in sensory abilities and the associated cognitive abilities have been key features of animal evolution [138,139]. This suggests the ability to experience pain may have arisen multiple times, even within a single phylum such as Arthropoda [138,139]. If the above proposal is correct, then it might provide some guidance as to which species might experience pain due to their flexible decision-making following tissue damage [25]. Animals with a mobile predatory lifestyle require rapid decision-making and are thus expected to be the most likely candidates for sentience. These may be found among crustaceans, insects, centipedes, scorpions, and spiders, and these appear to be worthy of detailed investigations.

8. Conclusions

There is no proof of pain in any animal, but it is possible to determine if pain is possible or even probable by determining if the expected criteria are fulfilled [12]. Here, there has been a focus on the behavioural criteria linked to the function of pain, i.e., changes that help the animal to subsequently avoid the conditions that produced the pain and changes that should enhance healing and survival. Such activities are found in crustaceans, insects, and, to a lesser extent, spiders. There are few such indicators in centipedes, millipedes, horseshoe crabs, and scorpions but these groups have received little attention regarding possible pain.

Pain offers advantages beyond those gained from nociceptive reflexes [8,15] but there are costs. These involve the development of neural circuits that enable pain to function. Further costs may occur, however, if animals manipulate the pain systems of others for their own gain. For example, males might evolve genitals that damage the reproductive tract of females, possibly to dissuade the females from mating with another male [82]. Females may also dissuade males from prolonged copulation [86]. Further, animals engage in fights over resources, and it is possible that pain is inflicted to encourage the opponent to give up. It is known that animals trade-off the avoidance of noxious stimuli with competing motivations and this might provide the mechanism by which animals incur greater costs in fights for high-value resources [90,91]. Finally, many arthropods produce venom, especially venom that dissuades a potential predator (or competitor). These venoms are often painful to humans, but it seems likely that at least some components of defensive venoms arose to function against arthropods. That suggests they might be effective at triggering pain responses in other arthropods [105]. These situations of intraspecific and interspecific conflict may also provide insights into the evolution of pain.

High cognitive ability has been suggested to be a requirement for pain [111]. Cognitive ability appears to increase as selection acts to improve the sensory systems and thus increase the information-gathering abilities of animals. The cognitive abilities of some arthropods are surprising and include the integration of information gained by diverse sensory systems and possible problem-solving in hermit crabs [90]. Also of note are concept formation [131] and social learning [129] in bees. This ability to integrate and manipulate information leads to a wider choice of behavioural responses to tissue damage and those choices might be based on some expectation of the utility of each candidate [25]. It has been argued that such an ability might result in the pain experience to guide and motivate the animal as it attempts to cope with tissue damage and make the best of a difficult situation [25]. There is evidence to suggest this may have occurred in some groups of arthropods, but probably not all.

Funding: This research received no external funding.

Institutional Review Board Statement: Not applicable.

Informed Consent Statement: Not applicable.

Data Availability Statement: Not applicable.

Conflicts of Interest: The author declares no conflict of interest.

References

1. Giribet, G.; Edgecombe, G.D. The phylogeny and evolutionary history of arthropods. *Curr. Biol.* **2019**, *29*, R592–R602.2. [CrossRef]
2. Zhang, J.; Walker, W.B.; Wang, G. Pheromone reception in moths: From molecules to behaviors. *Prog. Mol. Biol. Transl. Sci.* **2015**, *130*, 109–128. [CrossRef]
3. Barr, S.; Laming, P.R.; Dick, J.T.; Elwood, R.W. Nociception or pain in a decapod crustacean? *Anim. Behav.* **2008**, *75*, 745–751. [CrossRef]
4. Puri, S.; Faulkes, Z. Can crayfish take the heat? *Procambarus clarkii* show nociceptive behaviour to high temperature stimuli, but not low temperature or chemical stimuli. *Biol. Open* **2015**, *4*, 441–448. [CrossRef]
5. Gibbons, M.; Chittka, L. A framework for evaluating evidence of pain in animals. *Anim. Sentience* **2022**, *7*, 28. [CrossRef]
6. Raja, S.N.; Carr, D.B.; Cohen, M.; Finnerup, N.B.; Flor, H.; Gibson, S.; Keefe, F.J.; Mogil, J.S.; Ringkamp, M.; Sluka, K.A.; et al. The revised International Association for the Study of Pain definition of pain: Concepts, challenges, and compromises. *Pain* **2020**, *161*, 1976–1982. [CrossRef]
7. Crook, R.J. Behavioral and neurophysiological evidence suggests affective pain experience in octopus. *iScience* **2021**, *24*, 102229. [CrossRef]
8. Elwood, R.W. Pain and Suffering in Invertebrates? *ILAR J.* **2011**, *52*, 175–184. [CrossRef]
9. Crump, A.; Browning, H.; Schnell, A.; Burn, C.; Birch, J. Sentience in decapod crustaceans: A general framework and review of the evidence. *Anim. Sentience* **2022**, *7*, 1. [CrossRef]
10. Diggles, B.K. Food for thought: Review of some scientific issues related to crustacean welfare. *ICES J. Mar. Sci.* **2019**, *76*, 66–81. [CrossRef]
11. Adamo, S. Is it pain if it does not hurt? On the unlikelihood of insect pain. *Can. Entomol.* **2019**, *151*, 685–695. [CrossRef]
12. Bateson, P. Assessment of pain in animals. *Anim. Behav.* **1991**, *42*, 827–839. [CrossRef]
13. Smith, J.A.; Boyd, K.M. *Lives in the Balance: The Ethics of Using Animals in Biomedical Research*; The Report of a Working Party of the Institute of Medical Ethics; Oxford University Press: Oxford, UK, 1991. [CrossRef]
14. Birch, J.; Burn, C.; Schnell, A.K.; Browning, H.; Crump, A. *Review of the Evidence of Sentience in Cephalopod Molluscs and Decapod Crustaceans*; Department for Environment, Food & Rural Affairs (Defra): London, UK, 2021.
15. Sneddon, L.U.; Elwood, R.W.; Adamo, S.A.; Leach, M.C. Defining and assessing animal pain. *Anim. Behav.* **2014**, *97*, 201–212. [CrossRef]
16. Elwood, R.W. Pros and cons of a framework for evaluating potential pain in decapods. *Anim. Sentience* **2022**, *7*, 29. [CrossRef]
17. Irvine, E. Independence, weight and priority of evidence for sentience. *Anim. Sentience* **2022**, *32*, 423. [CrossRef]
18. Jablonka, E.; Ginsburg, S. Pain sentience criteria and their grading. *Anim. Sentience* **2022**, *7*, 4. [CrossRef]
19. Crook, R.J.; Dickson, K.; Hanlon, R.T.; Walters, E.T. (2014) Nociceptive sensitization reduces predation risk. *Curr. Biol.* **2014**, *24*, 1121–1125. [CrossRef]
20. Seymour, B.; Crook, R.J.; Chen, Z.S. Post-injury pain and behaviour: A control theory perspective. *Nat. Rev. Neurosci.* **2023**, *24*, 378–392. [CrossRef]
21. Appel, M.; Elwood, R.W. Motivational trade-offs and potential pain experience in hermit crabs. *Appl. Anim. Behav. Sci.* **2009**, *119*, 120–124. [CrossRef]
22. Elwood, R.W.; Appel, M. Pain experience in hermit crabs? *Anim. Behav.* **2009**, *77*, 1243–1246. [CrossRef]
23. Sneddon, L.U. Evolution of nociception in vertebrates: Comparative analysis of lower vertebrates. *Brain Res. Rev.* **2004**, *46*, 123–130. [CrossRef] [PubMed]
24. Gibbons, M.; Crump, A.; Barrett, E.; Sarlak, S.; Birch, J.; Chittka, L. Can insects feel pain? A review of the neural and behavioural evidence. *Adv. Insect Physiol.* **2022**, *63*, 155–229. [CrossRef]
25. Farnsworth, K.D.; Elwood, R.W. Why it hurts: With freedom comes the biological need for pain. *Anim. Cogn.* **2023**, *26*, 1259–1275. [CrossRef] [PubMed]
26. Okada, S.; Hirano, N.; Abe, T.; Nagayama, T. Aversive operant conditioning alters the phototactic orientation of the marbled crayfish. *J. Exp. Biol.* **2021**, *224 Pt 6*, jeb242180. [CrossRef] [PubMed]
27. Magee, B.; Elwood, R.W. Shock avoidance by discrimination learning in the shore crab (*Carcinus maenas*) is consistent with a key criterion for pain. *J. Exp. Biol.* **2013**, *216*, 353–358. [CrossRef]
28. Magee, B.T.; Elwood, R.W. No discrimination shock avoidance with sequential presentation of stimuli but shore crabs still reduce shock exposure. *Biol. Open* **2016**, *5*, 883–888. [CrossRef]
29. Dyer, A.G.; Neumeyer, C. Simultaneous and successive colour discrimination in the honeybee (*Apis mellifera*). *J. Comp. Physiol. A* **2005**, *191*, 547–557. [CrossRef]
30. Fossat, P.; Bacqué-Cazenave, J.; De Deurwaerdère, P.; Delbecque, J.-P.; Cattaert, D. Anxiety-like behavior in crayfish is controlled by serotonin. *Science* **2014**, *344*, 1293–1297. [CrossRef]
31. Fossat, P.; Bacqué-Cazenave, J.; De Deurwaerdere, P.; Cattaert, D.; Delbecque, J.-P. Serotonin, but not dopamine, controls the stress response and anxiety-like behavior in the crayfish Procambarus clarkii. *J. Exp. Biol.* **2015**, *218*, 2745–2752. [CrossRef]
32. Maza, F.J.; Urbano, F.J.; Delorenzi, A. Aversive memory conditioning induces fluoxetine-dependent anxiety-like states in the crab *Neohelice ranulate*. *J. Exp. Biol.* **2023**, *226*, jeb245590. [CrossRef]
33. Perrot-Minnot, M.-J.; Banchetry, L.; Cézilly, F. Anxiety-like behaviour increases safety from fish predation in an amphipod crustacea. *R. Soc. Open Sci.* **2017**, *4*, 171558. [CrossRef] [PubMed]

34. Elwood, R.W.; Stewart, A. The timing of decisions during shell investigation by the hermit crab, *Pagurus bernhardus*. *Anim. Behav.* **1985**, *33*, 620–627. [CrossRef]
35. Elwood, R.W. Motivational change during resource assessment by hermit crabs. *J. Exp. Mar. Biol. Ecol.* **1995**, *193*, 41–55. [CrossRef]
36. Appel, M.; Elwood, R.W. Gender differences, responsiveness and memory of a potentially painful event in hermit crabs. *Anim. Behav.* **2009**, *78*, 1373–1379. [CrossRef]
37. Elwood, R.W. Discrimination between nociceptive reflexes and more complex responses consistent with pain in crustaceans. *Philos. Trans. R. Soc. B Biol. Sci.* **2019**, *374*, 20190368. [CrossRef]
38. Magee, B.; Elwood, R.W. Trade-offs between predator avoidance and electric shock avoidance in hermit crabs demonstrate a non-reflexive response to noxious stimuli consistent with prediction of pain. *Behav. Process.* **2016**, *130*, 31–35. [CrossRef]
39. Elwood, R.W.; Dalton, N.; Riddell, G. Aversive responses by shore crabs to acetic acid but not to capsaicin. *Behav. Process.* **2017**, *140*, 1–5. [CrossRef]
40. McCambridge, C.; Dick, J.T.A.; Elwood, R.W. Effects of autotomy compared to manual declawing on contests between males for females in the edible crab *Cancer pagurus*: Implications for fishery practice and animal welfare. *J. Shellfish. Res.* **2016**, *35*, 1037–1044. [CrossRef]
41. Dyuizen, I.V.; Kotsyuba, E.P.; Lamash, N.E. Changes in the nitric oxide system in the shore crab *Hemigrapsus sanguineus* (Crustacea, decapoda) CNS induced by a nociceptive stimulus. *J. Exp. Biol.* **2012**, *215*, 2668–2676. [CrossRef] [PubMed]
42. Patterson, L.; Dick, J.T.A.; Elwood, R.W. Physiological stress responses in the edible crab, *Cancer pagurus*, to the fishery practice of de-clawing. *Mar. Biol.* **2007**, *152*, 265–272. [CrossRef]
43. Fiorito, G. Is there 'pain' in invertebrates? *Behav. Process.* **1986**, *12*, 383–388. [CrossRef]
44. Barr, S.; Elwood, R.W. No evidence of morphine analgesia to noxious shock in the shore crab, *Carcinus maenas*. *Behav. Process.* **2011**, *86*, 340–344. [CrossRef]
45. Pitman, J.L.; DasGupta, S.; Krashes, M.J.; Leung, B.; Paola, N.; Perrat, P.N.; Waddell, S. There are many ways to train a fly. *Fly* **2009**, *3*, 3–9. [CrossRef]
46. Tully, T.; Quinn, W.G. Classical conditioning and retention in normal and mutant *Drosophila melanogaster*. *J. Comp. Physiol. A* **1985**, *157*, 263–277. [CrossRef]
47. Barraco, D.A.; Lovell, K.L.; Eisenstein, E.M. Effects of cycloheximide and puromycin on learning and retention in the cockroach, *P. americana*. *Pharmacol. Biochem. Behav.* **1981**, *15*, 489–494. [CrossRef]
48. Abramson, C.I.; Morris, A.W.; Michaluk, L.M.; Squire, J. Antistatic foam as a shocking surface for behavioral studies with honeybees (Hymenoptera: Apidae) and American cockroaches (Orthoptera: Blattelidae). *J. Entomol. Sci.* **2004**, *39*, 562–566.
49. Dylla, K.V.; Raiser, G.; Galizia, C.G.; Szyszka, P. Trace conditioning in *Drosophila* induces associative plasticity in mushroom body Kenyon cells and dopaminergic neurons. *Front. Neural Circuits* **2017**, *11*, 42. [CrossRef] [PubMed]
50. Groening, J.; Venini, D.; Srinivasan, M.V. In search of evidence for the experience of pain in honeybees: A self-administration study. *Sci. Rep.* **2017**, *7*, 45825. [CrossRef] [PubMed]
51. Gibbons, M.; Sarlak, S.; Chittka, L. Descending control of nociception in insects? *Proc. R. Soc. B.* **2022**, *289*, 20220599. [CrossRef]
52. Bateson, M.; Desire, S.; Gartside, S.E.; Wright, G.A. Agitated honeybees exhibit pessimistic cognitive biases. *Curr. Biol.* **2011**, *21*, 1070–1073. [CrossRef]
53. Ings, T.C.; Chittka, L. Speed-accuracy tradeoffs and false alarms in bee responses to cryptic predators. *Curr. Biol.* **2008**, *18*, 1520–1524. [CrossRef] [PubMed]
54. Khuong, T.M.; Wang, Q.-P.; Manion, J.; Oyston, L.J.; Lau, M.-T.; Towler, H.; Lin, Y.Q.; Neely, G.G. Nerve injury drives a heightened state of vigilance and neuropathic sensitization in *Drosophila*. *Sci. Adv.* **2019**, *5*, eaaw4099. [CrossRef]
55. Gibbons, M.; Versace, E.; Andrew Crump, A.; Baran, B.; Chittka, L. Motivational trade-offs and modulation of nociception in bumblebees. *Proc. Natl. Acad. Sci. USA* **2022**, *119*, e2205821119. [CrossRef]
56. Gibbons, M. Neural and Behavioural Indicators of Pain in Insects. Ph.D. Thesis, Queen Mary University of London, London, UK, 2023.
57. Emberts, Z.; St. Mary, C.M.; Howard, C.C.; Forthman, M.; Bateman, P.W.; Somjee, U.; Hwang, W.S.; Li, D.; Kimball, R.T.; Miller, C.W. The evolution of autotomy in leaf-footed bugs. *Evolution* **2020**, *74*, 897–910. [CrossRef] [PubMed]
58. Maginnis, T.L. Autotomy in a stick insect (Insecta: Phasmida): Predation versus molting. *Fla. Entomol.* **2008**, *91*, 126–127. [CrossRef]
59. Mikhalevich, I.; Powell, R. Minds without spines: Evolutionarily inclusive animal ethics. *Anim. Sentience* **2020**, *5*, 1. [CrossRef]
60. Elwood, R.W. Do arthropods respond to noxious stimuli purely by reflex? *Anim. Sentience* **2020**, *5*, 10. [CrossRef]
61. Dunoyer, L.A.; Seifert, A.W.; Van Cleve, J. Evolutionary bedfellows: Reconstructing the ancestral state of autotomy and regeneration. *J. Exp. Zool. B Mol. Dev. Evol.* **2021**, *336*, 94–115. [CrossRef] [PubMed]
62. Punzo, F. Leg autotomy and avoidance behavior in response to a predator in the wolf spider, *Schizocosa avida* (Araneae, Lycosidae). *J. Arachnol.* **1997**, *25*, 202–205.
63. Nakamura, T.; Yamashita, S. Learning and discrimination of colored papers in jumping spiders (Araneae, Salticidae). *J. Comp. Physiol. A* **2000**, *186*, 897–901. [CrossRef] [PubMed]
64. Bednarski, J.V.; Taylor, P.; Jakob, E.M. Optical cues used in predation by jumping spiders, *Phidippus audax* (Araneae, Salticidae). *Anim. Behav.* **2012**, *84*, 1221–1227. [CrossRef]

65. Peckmezian, T.; Taylor, P.W. Electric shock for aversion training of jumping spiders: Towards an arachnid model of avoidance learning. *Behav. Process.* **2015**, *113*, 99–104. [CrossRef]
66. Kralj-Fišer, S.; Gregorič, M. Spider Welfare. In *The Welfare of Invertebrate Animals. Animal Welfare*; Carere, C., Mather, J., Eds.; Springer: Cham, Switzerland, 2019; Volume 18, pp. 105–122. [CrossRef]
67. Eisner, T.; Camazine, S. Spider leg autotomy induced by prey venom injection: An adaptive response to "pain"? *Proc. Natl. Acad. Sci. USA* **1983**, *80*, 3382–3385. [CrossRef]
68. Punzo, F. Habituation, avoidance learning, and spatial learning in the giant whipscorpion, *Mastigoproctus giganteus* (Lucas) (Arachnida, Uropygi). *Bull. Br. Arachnol. Soc.* **2005**, *13*, 138–144.
69. Gorman, R. Atlantic horseshoe crabs and endotoxin testing: Perspectives on alternatives, sustainable methods, and the 3Rs (Replacement, Reduction and Refinement). *Front. Mar. Sci.* **2020**, *7*, 582132. [CrossRef]
70. Anderson, R.L.; Watson, W.H.; Chabot, C.C. Sublethal behavioral and physiological effects of the biomedical bleeding process on the American horseshoe crab, *Limulus polyphemus*. *Biol. Bull.* **2013**, *225*, 137–151. [CrossRef]
71. Owings, M.; Chabot, C.; Watson, W. Effects of the biomedical bleeding process on the behavior and hemocyanin levels of the American horseshoe crab (*Limulus polyphemus*). *Fish.Bull.* **2020**, *118*, 225–239. [CrossRef]
72. Makous, W.L. Conditioning in the horseshoe crab. *Psychon. Sci.* **2013**, *14*, 4–5. [CrossRef]
73. Barron, A.B.; Klein, C. What insects can tell us about the origins of consciousness. *Proc. Natl. Acad. Sci. USA* **2016**, *113*, 4900–4908. [CrossRef] [PubMed]
74. Sandeman, D.C.; Kenning, M.; Harzsch, S. Adaptive trends in malacostracan brain form and function related to behavior. In *Crustacean Nervous System and Their Control of Behaviour*; Derby, C., Thiel, M., Eds.; Oxford University Press: Oxford, UK, 2014; pp. 11–48.
75. Strausfeld, N.J.; Wolff, G.H.; Sayre, M.E. Mushroom body evolution demonstrates homology and divergence across Pancrustacea. *eLife* **2020**, *9*, e52411. [CrossRef]
76. Webster, S.G. Measurement of crustacean hyperglycaemic hormone levels in the edible crab *Cancer pagurus* during emersion stress. *J. Exp. Biol.* **1996**, *199*, 1579–1585. [CrossRef] [PubMed]
77. Even, N.; Devaud, J.-M.; Barron, A.B. General stress responses in the honeybee. *Insects* **2012**, *3*, 1271–1298. [CrossRef]
78. Stentiford, G.D.; Chang, E.S.; Chang, S.A.; Neil, D.M. Carbohydrate dynamics and the crustacean hyperglycemic hormone (CHH): eVects of parasitic infection in Norway lobsters (*Nephrops norvegicus*). *Gen. Comp. Endocrinol.* **2001**, *121*, 13–22. [CrossRef]
79. Elwood, R.W.; Adams, L. Electric shock causes physiological stress responses in shore crabs, consistent with prediction of pain. *Biol. Lett.* **2015**, *11*, 20150800. [CrossRef]
80. Lessells, C.M. The evolutionary outcome of sexual conflict. *Philos. Trans. R. Soc. B: Biol. Sci.* **2006**, *361*, 301–317. [CrossRef] [PubMed]
81. Johnstone, R.A.; Keller, L. How males can gain by harming their mates: Sexual conflict, seminal toxins, and the cost of mating. *Am. Nat.* **2000**, *156*, 368–377. [CrossRef]
82. Crudgington, H.; Siva-Jothy, M. Genital damage, kicking and early death. *Nature* **2000**, *407*, 855–856. [CrossRef] [PubMed]
83. Ma, N.; Gong, D.; Mao, A.; Zhao, Y.; Jiao, X.; Liu, J.; Peng, Y.; Zhang, S. Traumatic mating causes strict monandry in a wolf spider. *Zool Res.* **2023**, *44*, 101–104. [CrossRef]
84. Simone, Y.; van der Meijden, A. Armed stem to stinger: A review of the ecological roles of scorpion weapons. *J. Venom. Anim. Toxins Incl. Trop. Dis.* **2021**, *27*, e20210002. [CrossRef] [PubMed]
85. Sentenská, L.; Graber, F.; Richard, M.; Kropf, C. Sexual dimorphism in venom gland morphology in a sexually stinging scorpion. *Biol. J. Linn. Soc.* **2017**, *122*, 429–443. [CrossRef]
86. Rivas-Torres, A.; Di Pietro, V.; Cordero-Rivera, A. Sex wars: A female genital spine forces male damselflies to shorten copulation duration. *Evolution* **2023**, *77*, 1659–1666. [CrossRef] [PubMed]
87. Crump, A.; Bethell, E.J.; Earley, R.; Lee, V.E.; Mendl, M.; Oldham, L.; Turner, S.P.; Arnott, G. Emotion in animal contests. *Proc. Roy. Soc. B* **2020**, *287*, 20201715. [CrossRef]
88. Green, P.A.; Patek, S.N. Mutual assessment during ritualized fighting in mantis shrimp (Stomatopoda). *Proc. R. Soc. B* **2018**, *285*, 20172542. [CrossRef]
89. Briffa, M.; Elwood, R.W. The power of shell rapping influences rates of eviction in hermit crabs. *Behav. Ecol.* **2000**, *11*, 288–293. [CrossRef]
90. Elwood, R.W. Hermit crabs, shells, and sentience. *Anim. Cogn.* **2022**, *25*, 1241–1257. [CrossRef] [PubMed]
91. Arnott, G.; Elwood, R.W. Information-gathering and decision-making about resource value in animal contests. *Anim. Behav.* **2008**, *76*, 529–542. [CrossRef]
92. LeBrun, E.G.; Jones, N.T.; Gilbert, L.E. Chemical warfare among invaders: A detoxification interaction facilitates an ant invasion. *Science* **2014**, *343*, 1014–1017. [CrossRef]
93. Dashevsky, D.; Baumann, K.; Undheim, E.A.B.; Nouwens, A.; Ikonomopoulou, M.P.; Schmidt, J.O.; Ge, L.; Kwok, H.F.; Rodriguez, J.; Fry, B.G. Functional and Proteomic Insights into Aculeata Venoms. *Toxins* **2023**, *15*, 224. [CrossRef]
94. Branstetter, M.G.; Danforth, B.N.; Pitts, J.P.; Faircloth, B.C.; Ward, P.S.; Buffington, M.L.; Gates, M.W.; Kula, R.R.; Brady, S.G. Phylogenomic insights into the evolution of stinging wasps and the origins of ants and bees. *Curr. Biol.* **2017**, *27*, 1019–1025. [CrossRef]

95. Koludarov, I.; Velasque, M.; Timm, T.; Greve, C.; Hamadou, A.B.; Gupta, D.K.; Lochnit, G.; Heinzinger, M.; Vilcinskas, A.; Gloag, R.; et al. A common venomous ancestor? Prevalent bee venom genes evolved before the aculeate stinger while few major toxins are bee-specific. *bioRxiv* **2022**. [CrossRef]
96. Evans, E.R.J.; Northfield, T.D.; Norelle, L.; Daly, N.L.; Wilson, D.T. Venom costs and optimization in scorpions. *Front. Ecol. Evol.* **2019**, *7*, 196. [CrossRef]
97. Niermann, C.N.; Tate, T.G.; Suto, A.L.; Barajas, R.; White, H.A.; Guswiler, O.D.; Secor, S.M.; Rowe, A.H.; Rowe, M.P. Defensive Venoms: Is Pain Sufficient for Predator Deterrence? *Toxins* **2020**, *12*, 260. [CrossRef]
98. Walker, A.A.; Mayhew, M.L.; Jin, J.; Herzig, V.; Undheim, E.A.; Sombke, A.; Fry, B.G.; Meritt, D.J.; King, G.F. The assassin bug *Pristhesancus plagipennis* produces two distinct venoms in separate gland lumens. *Nat. Commun.* **2018**, *9*, 755. [CrossRef]
99. Rügen, N.; Jenkins, T.P.; Wielsch, N.; Vogel, H.; Hempel, B.-F.; Süssmuth, R.D.; Ainsworth, S.; Cabezas-Cruz, A.; Vilcinskas, A.; Tonk, M. Hexapod assassins' potion: Venom composition and bioactivity from the Eurasian assassin bug *Rhynocoris iracundus*. *Biomedicines* **2021**, *9*, 819. [CrossRef]
100. Diochot, S. Pain-related toxins in scorpion and spider venoms: A face to face with ion channels. *J. Venom. Anim. Toxins incl. Trop. Dis.* **2021**, *27*, e20210026. [CrossRef]
101. Herzig, V.; Sunagar, K.; Wilson, D.T.R.; Pineda, S.S.; Israel, M.R.; Dutertre, S.; McFarland, B.S.; Undheim, E.A.B.; Hodgson, W.C.; Alewood, P.F.; et al. Australian funnel-web spiders evolved human-lethal δ-hexatoxins for defense against vertebrate predators. *Proc. Natl. Acad. Sci. USA* **2020**, *117*, 24920–24928. [CrossRef]
102. Eisner, T.; Aneshansley, D.; del Campo, M.L.; Eisner, M.; Frank, J.H.; Deyrup, M. Effect of bombardier beetle spray on a wolf spider: Repellency and leg autotomy. *Chemoecology* **2006**, *16*, 185–189. [CrossRef]
103. Sugiura, S. Beetle bombing always deters praying mantises. *PeerJ* **2021**, *9*, e11657. [CrossRef]
104. Eisner, T.; Aneshansley, D.J. Spray aiming in the bombardier beetle: Photographic evidence. *Proc. Natl. Acad. Sci. USA* **1999**, *96*, 9705–9709. [CrossRef]
105. Murphy, S.M.; Leahy, S.M.; Williams, L.S.; Lill, J.T. Stinging spines protect slug caterpillars (Limacodidae) from multiple generalist predators. *Behav. Ecol.* **2010**, *21*, 153–160. [CrossRef]
106. Braithwaite, V.A. *Do Fish Feel Pain?* Oxford University Press: Oxford, UK, 2010.
107. Chandroo, K.P.; Duncan, I.J.H.; Moccia, R.D. Can fish suffer? Perspectives on sentience, pain, fear and stress. *Appl. Anim. Behav. Sci.* **2004**, *86*, 225–250. [CrossRef]
108. Duncan, I.J.H. Animal welfare defined in terms of feelings. *Acta. Agric. Scand. A Suppl.* **1996**, *27*, 29–35.
109. Duncan, I.J.H.; Petherick, C. The implications of cognitive processes for animal welfare. *J. Anim. Sci.* **1991**, *69*, 5017–5022. [CrossRef]
110. Dawkins, M.S. Through animal eyes: What behaviour tells us. *Appl. Anim. Behav. Sci.* **2006**, *100*, 4–10. [CrossRef]
111. Broom, D.M. Cognitive ability and sentience: Which aquatic animals should be protected? *Dis. Aquat. Org.* **2007**, *75*, 99–108. [CrossRef]
112. Dukas, R. Constraints on Information Processing and Their Effects on Behavior. In *Cognitive Ecology*; Dukas, R., Ed.; University of Chicago Press: Chicago, IL, USA, 1998; pp. 89–127.
113. Jackson, N.W.; Elwood, R.W. How animals make assessments: Information gathering by the hermit crab, *Pagurus bernhardus*. *Anim. Behav.* **1989**, *38*, 951–957. [CrossRef]
114. Krieger, J.; Hörnig, M.K.; Laidre, M.E. Shells as 'extended architecture': To escape isolation, social hermit crabs choose shells with the right external architecture. *Anim. Cogn.* **2020**, *23*, 1177–1187. [CrossRef]
115. Sonoda, K.; Asakura, A.; Minoura, M.; Elwood, R.W.; Gunji, P. Hermit crabs perceive the extent of their virtual bodies. *Biol. Lett.* **2012**, *8*, 495–497. [CrossRef]
116. Elwood, R.W.; Pothanikat, E.; Briffa, M. Honest and dishonest displays, motivational state, and subsequent decisions in hermit crab shell fights. *Anim. Behav.* **2006**, *72*, 853–859. [CrossRef]
117. Arnott, G.; Elwood, R.W. Fighting for shells: How private information about resource value changes hermit crab pre-fight displays and escalated fight behaviour. *Proc. Roy. Soc. B* **2007**, *274*, 3011–3017. [CrossRef]
118. Dowds, B.M.; Elwood, R.W. Shell wars: Assessment strategies and the timing of decisions in hermit crab fights. *Behaviour* **1983**, *85*, 1–24. Available online: https://psycnet.apa.org/doi/10.1163/156853983X00011 (accessed on 7 August 2023). [CrossRef]
119. Briffa, M.; Elwood, R.W.; Dick, J.T.A. Analyses of repeated signals during hermit crab shell fights. *Proc. Roy. Soc. B* **1998**, *265*, 1467–1474. [CrossRef]
120. Briffa, M.; Elwood, R.W. Decision rules, energy metabolism and vigour in hermit crab fights. *Proc. Roy. Soc. B* **2001**, *268*, 1841–1847. [CrossRef]
121. Briffa, M.; Elwood, R.W. Rapid change in energetic status in fighting animals: Causes and effects of strategic decisions. *Anim. Behav.* **2005**, *70*, 119–124. [CrossRef]
122. Briffa, M.; Elwood, R.W. Use of energy reserves in fighting hermit crabs. *Proc. Roy. Soc. B* **2004**, *271*, 373–379. [CrossRef]
123. Gherardi, F.; Atema, J. Memory of social partners in hermit crab dominance. *Ethology* **2005**, *111*, 271–285. [CrossRef]
124. Hazlett, B.A. Assessments during shell exchanges by the hermit crab *Clibanarius vittatus*: The complete negotiator. *Anim. Behav.* **1996**, *51*, 567–573. [CrossRef]
125. Vannini, M.; Cannicci, S. Homing behaviour and possible cognitive maps in crustacean decapods. *J. Exp. Mar. Biol. Ecol.* **1995**, *193*, 67–91. [CrossRef]

126. von Hagen, H. Nachweis einer kinasthetischen Orientterung bei *Uca rapax*. *Z. Morphol. Okol. Tiere* **1967**, *58*, 301–320. [CrossRef]
127. Lohmann, K.J.; Pentcheff, N.D.; Nevitt, G.A.; Stetten, G.D.; Zimmerfaust, R.K.; Jarrard, H.E.; Boles, L.C. Magnetic orientation of spiny lobsters in the ocean: Experiments with undersea coil systems. *J. Exp. Biol.* **1995**, *198*, 2041–2048. [CrossRef]
128. Boles, L.C.; Lohmann, K.J. True navigation in spiny lobsters. *Nature* **2003**, *421*, 60–63. [CrossRef] [PubMed]
129. Chittka, L. *The Mind of a Bee*; Princeton University Press: Princeton, NJ, USA, 2022.
130. Benard, J.; Stach, S.; Giurfa, M. Categorisation of visual stimuli in the honeybee *Apis mellifera*. *Anim. Cogn.* **2006**, *9*, 257–270. [CrossRef] [PubMed]
131. Giurfa, M.; Eichmann, B.; Menzel, R. Symmetry perception in an insect. *Nature* **1996**, *382*, 458–461. [CrossRef]
132. Alem, S.; Perry, C.J.; Zhu, X.; Loukola, O.J.; Ingraham, T.; Søvik, E.; Chittka, L. Associative mechanisms allow for social learning and cultural transmission of string pulling in an insect. *PLoS Biol.* **2016**, *14*, e1002564. [CrossRef] [PubMed]
133. Bartos, M. Alternative predatory tactics in a juvenile hunting spider. *J. Arachnol.* **2008**, *36*, 300–305. [CrossRef]
134. Jackson, R.R.; Pollard, S.D.; Li, D.; Fijn, N. Interpopulation variation in the risk-related decisions of *Portia labiata*, an araneophagic jumping spider (Araneae, Salticidae), during predatory sequences with spitting spiders. *Anim. Cogn.* **2002**, *5*, 215–223. [CrossRef] [PubMed]
135. Tarsitano, M.S. Route selection by a jumping spider (*Portia labiata*) during the locomotory phase of a detour. *Anim. Behav.* **2006**, *72*, 1437–1442. [CrossRef]
136. Sherwin, C.M. Can invertebrates suffer? Or how robust is argument-by-analogy? *Anim. Welf.* **2001**, *10*, 103–108. [CrossRef]
137. Edgecombe, G.D.; Legg, D.A. Origins and early evolution of arthropods. *Palaeontology* **2014**, *57*, 457–468. [CrossRef]
138. Godfrey-Smith, P. *Metazoa: Animal Minds and the Birth of Consciousness*; William Collins: London, UK, 2020.
139. Lacalli, T. An evolutionary perspective on chordate brain organization and function: Insights from amphioxus, and the problem of sentience. *Phil. Trans. R. Soc. B* **2022**, *377*, 20200520. [CrossRef] [PubMed]

animals

MDPI

Article

The Use of Isoflurane and Adjunctive Magnesium Chloride Provides Fast, Effective Anaesthetization of *Octopus vulgaris*

Anna Di Cosmo [1,2,*], Valeria Maselli [1], Emanuela Cirillo [1,2], Mariangela Norcia [1], Heethaka K. S. de Zoysa [1,3], Gianluca Polese [1] and William Winlow [1,4,*]

1 Department of Biology, University of Naples Federico II, 80126 Naples, Italy; valeria.maselli@unina.it (V.M.); emanuela.cirillo@unina.it (E.C.); mariangela.norcia@unina.it (M.N.); krishanthasameeradezoysa.heethaka@unina.it (H.K.S.d.Z.); gianluca.polese@unina.it (G.P.)

2 PNRR "MNESYS", University of Naples Federico II, 80126 Naples, Italy

3 Department of Bioprocess Technology, Faculty of Technology, Rajarata University of Sri Lanka, Mihintale 50300, Sri Lanka

4 Institute of Ageing and Chronic Diseases, University of Liverpool, Liverpool L69 3BX, UK

* Correspondence: dicosmo@unina.it (A.D.C.); bill.winlow@gmail.com (W.W.)

Simple Summary: Anaesthetising invertebrates is a welfare issue, but because there are so many different invertebrates (97% of all animal species), it is of the utmost importance to work out the most appropriate way to anaesthetise each group or even particular species of animals. In order to reduce stress and pain to the animals, clinical anaesthetics offer a standardised approach, given their known and well-studied effects at both system and cellular levels. Here, we show that the clinical anaesthetic isoflurane (1%), often used by veterinarians, can anaesthetise octopuses quickly and efficiently when used in conjunction with 1% magnesium chloride, which acts as a muscle relaxant. Recovery is equally quick, and the whole process from anaesthetisation to recovery is completed in 30 to 35 min, minimising animal stress.

Abstract: A wide variety of substances have been used to anaesthetise invertebrates, but many are not anaesthetics and merely incapacitate animals rather than preventing pain. In essence, the role of an ideal general anaesthetic is to act as a muscle relaxant, an analgesic, an anaesthetic, and an amnesic. To achieve all these properties with a single substance is difficult, and various adjuvants usually need to be administered, resulting in a cocktail of drugs. In a clinical setting, the vast majority of patients are unaware of surgery being carried out and have no memory of it, so they can claim to have felt no pain, but this is much more difficult to demonstrate in invertebrates. Here, we show that 1% $MgCl_2$, a muscle relaxant, is a useful adjuvant for the clinical anaesthetic isoflurane on *Octopus vulgaris* when applied alone in seawater for 10 min before the clinical anaesthetic. After this, full anaesthesia can be achieved in 5 min using 1% isoflurane insufflated into the saline still containing $MgCl_2$. Full recovery takes place rapidly in about 10 to 15 min. The depth of anaesthesia was monitored using changes in respiratory rate, chromatophore pattern, and withdrawal movements of the arms and siphon. This methodology reduces stress on the animal and minimises the quantity of anaesthetic used.

Keywords: general anaesthetics; *Octopus vulgaris*; isoflurane; magnesium chloride; depth of anaesthesia; stress reduction

1. Introduction

In the past, the welfare of invertebrates was largely unexplored, and little attention has been paid to it by many experimental biologists. According to Mather [1], "Consideration of welfare of other animals often is anthropocentric, focussing on mammals similar to humans". Fortunately, this view is gradually changing as it becomes clear that other groups of animals may be sentient and capable of feeling pain and suffering in consequence, particularly decapod crustaceans [2,3], cephalopod molluscs [4–6], and possibly some insects. This may also be true of other groups of animals, and it behoves us to treat all invertebrates with respect when carrying out experimental procedures upon them.

Many different substances have been suggested and used for anaesthetising invertebrates [7], some more successfully than others. Clinically used general anaesthetics have several functions, and many different types exist, either volatile or systemic anaesthetics. We know the general principles behind their use because of the vast amount of clinical experience with them. In essence, the role of an ideal general anaesthetic is to act as a muscle relaxant, an analgesic, an anaesthetic, and an amnesic. Achieving all these properties is difficult with a single substance, and various adjuvants usually need to be used, resulting in the administration of a cocktail of drugs. In a clinical setting, the vast majority of patients are unaware of surgery being carried out and have no memory of it, so they can claim to have felt no pain, but this is much more difficult to demonstrate in invertebrates. However, there is strong support for the view that cephalopods can feel pain [8–10]. For example, they avoid locations previously associated with noxious stimuli [11]. There is also accumulating evidence for pain in arthropods [12]. Thus, we need to exert great care when handling all invertebrates to reduce stress and pain to the animals, particularly when carrying out physiological experiments.

For some time, it has been suggested that magnesium chloride by itself can act as an anaesthetic agent in cephalopods (e.g., [13–15]), but its main use is as a muscle relaxant, which may render an animal immobile. However, we cannot ensure that the animal is pain-free, and there has been substantial controversy on this issue [7,16]. Magnesium chloride works by competing with calcium ions to prevent synaptic release of neurotransmitters in the periphery and does not usually gain access to the central nervous system [5]. In 2018, we suggested that $MgCl_2$ might be useful as an adjunct to anaesthesia [5], and here we explore this concept in more detail. There is a fuller discussion of the mode of action of magnesium salts as muscle relaxants elsewhere [5], but it should be noted that $MgCl_2$ may also be a mild central sedative analgesic [17,18]. However, in our view, volatile anaesthetics are preferable, and isoflurane has proved successful for anaesthetising *Octopus vulgaris* [5,19]. Isoflurane is a non-flammable halogenated ether compound used by both clinicians and veterinarians. When administered via the respiratory system, it has direct access to both the peripheral and central nervous systems.

Properties of general anaesthetics. Clinical anaesthetics, such as isoflurane and related compounds, have well-known systemic and cellular actions, which have been demonstrated on mammals and on gastropod molluscs such as *Lymnaea stagnalis* [19] and on *Octopus vulgaris* [20]. They include ion channels and voltage-gated channels for sodium, potassium, and calcium and tend to block synaptic transmission [21]. These are general properties and actions and should not be ignored by investigators working on invertebrates. Furthermore, Keltz and Mashhour [22] recently provided good evidence that a generalisable mechanistic framework for the actions of general anaesthetics is emerging from studies of a wide range of species "from *Paramecium* to primates" and includes *Drosophila*, *Caenorhabditis elegans*, gastropod molluscs, such as *Lymnaea stagnalis*, etc., as well as plants.

We recently developed a successful method for anaesthetising *Octopus vulgaris* using the clinical anaesthetic isoflurane [20] without any adjunctive agents, but this took approaching two hours from the start of anaesthetisation to full recovery. In these experiments, we gradually increased the concentration of isoflurane in seawater from 0.5% to 2.0% and found that at 1.0% isoflurane, the chromatophores started to flash in an uncontrolled manner, a sign of loss of central motor control of chromatophores in *Octopus vulgaris* [23]. After this, the animals became relaxed, unresponsive to touch stimuli, and anaesthetised [20]. Recovery and anaesthetisation were judged physiologically by the rate of respiratory pumping because most clinical anaesthetics depress minute ventilation in mammals [24]. Two behavioural tests were also used: local withdrawal responses of the arms and siphon and loss of chromatophore patterning. A shorter period of anaesthetisation should reduce both the stress on the animal and the quantity of anaesthetic used. Here, we report, for the first time in *Octopus vulgaris*, that the use of $MgCl_2$, as both a pre-anaesthetic agent and as an adjunct to anaesthesia, significantly reduces the time needed to anaesthetise the animal successfully and also promotes more rapid recovery.

2. Materials and Methods

Animals

Nine specimens of *Octopus vulgaris* (5 males and 4 females; body weight, 0.5–1.0 kg) were captured in the Bay of Naples, bought from the local fish market, and transported to Prof. Di Cosmo's cephalopod facility [25] at the Department of Biology, University of Naples Federico II (Naples, Italy). Our research was approved following the European Directive 2010/63 EU L276, the Italian DL. 4/03/2014, n°26, and the ethical principles of Reduction, Refinement, and Replacement (Project n°608/2016-PR-17/06/2016; protocol n°DGSAF 0022292-P-03/10/2017). The test specimens were acclimatised [26], and were fed with fresh fish (*Engraulis encrasicolus*) as described in Maselli [27]. All octopuses fully recovered from the experimental procedures; they were kept in Prof. Di Cosmo's animal facility and later employed in other experiments.

Anaesthesia setup—In our study, inhalational anaesthesia was performed as described by Polese et al. [20]. The seawater surrounding the animal was aerated using a power pump and regulated by a flowmeter (1.8 L/min) in series with an isoflurane vapouriser and delivered to the bath via an air stone as previously described [20]. Except for the delivery of the anaesthetic into the aquatic medium, this is a standard clinical technique that allows the isoflurane to be delivered to the animal with air as the carrier gas, thus ensuring that the animal remains fully oxygenated throughout the procedure. All inhalational anaesthetics are delivered into the animal's system via an air–water interface, the respiratory epithelium in both vertebrates and octopuses, but in this case, it was via an aquatic medium. The experimental tank was kept in an enclosure and vented externally. It contained 1600 mL of seawater and was kept at 18 °C throughout the experiments. Each animal was left to acclimatise in the experimental tank for 10 min before delivering the air and isoflurane gas mixture via the air stone.

Protocol for anaesthesia—After the acclimatisation period, the animals were first moved gently into an experimental tank containing 1% $MgCl_2$ in aerated seawater for a period of 10 min. Then, the anaesthetic was insufflated in the airflow for 5 min at 1% concentration, after which the animal was maintained in fresh, clean seawater until full recovery. To determine the depth of anaesthesia, we monitored the respiratory rate as judged by the frequency of respiratory pumping and behavioural changes in the chromatophore pattern and withdrawal of the arms and siphon. We recorded the responses with a web camera (GoPro Hero10, 4K video at 30 frames per second (fps) in full-screen; internal auto-focus lens system).

The respiratory rate was taken every minute just before the touch test was performed, to avoid any change due to stimulation. The animals had widely varying pre-control respiratory rates (24–45 cycles per minute) as judged by the pumping movements of the mantle; so, for analytical purposes, we normalised the respiratory rate, with 100% being the observed pre-control value [20].

Touch stimuli induced withdrawal of the arms and siphon, and we monitored the changes in chromatophore pattern throughout each experiment. The withdrawal responses were elicited by the gentle application of a blunt Plexiglas probe to the arms and siphon. For analytical purposes, we categorised the responses as strong, medium, weak, or abolished, and these were given numerical values of 6, 4, 2, and 0, respectively. Using ImageJ software 1.46r [28], we measured a spot 10 pixels in size from the interbrachial membrane on frames taken every 5 min throughout the experiments. The colour intensity of each spot was measured based on the additive red, green, blue (RGB) colour model whereby a zero intensity (value, 0) for each component gives the darkest colour (no light, considered to be black) and full intensity for each component gives white (value, 255).

Statistical analyses—A one-way analysis of variance (ANOVA) was used to assess the significance of differences between the data at time 0 and those at other times (R-CRAN, [29].

3. Results

3.1. Actions of 1% $MgCl_2$ on Animal Behaviour

Upon entry to the 1% $MgCl_2$ in aerated seawater, most animals struggled, but within three minutes of application of $MgCl_2$, the animals visibly started to relax (4.17% reduction in the median normalised respiratory rate), and by 5 min, all were pale in colour (Figure 1) with weak touch reflexes (Figure 2). After 10 min, the animals were all pale, unreactive, and immobile, with a 19.50% reduction in the median % normalised respiratory rate. Thus, the animals appeared to be paralysed but continued to breathe steadily (Figure 3).

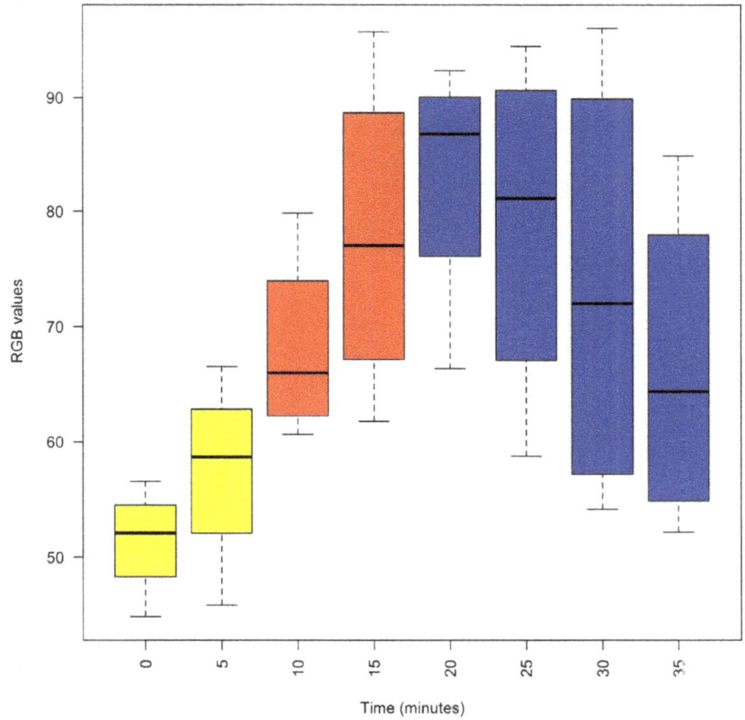

Figure 1. Box plot of the colour intensity of the interbrachial membrane as the concentration of isoflurane was progressively increased and then decreased. Yellow, 1% $MgCl_2$; Red, 1% $MgCl_2$ and 1% isoflurane; Blue, recovery in fresh aerated seawater. Colour intensity was measured using the RGB model whereby a zero intensity (value, 0) for each component gives the darkest colour (no light, considered to be black) and full intensity for each component gives white (value, 255). Thus, the higher the recorded value, the paler the interbrachial membrane and vice versa.

None showed flashing reactions due to chromatophore activation at any stage, presumably due to the inactivation of transmitter release from chromatophore motor neurons.

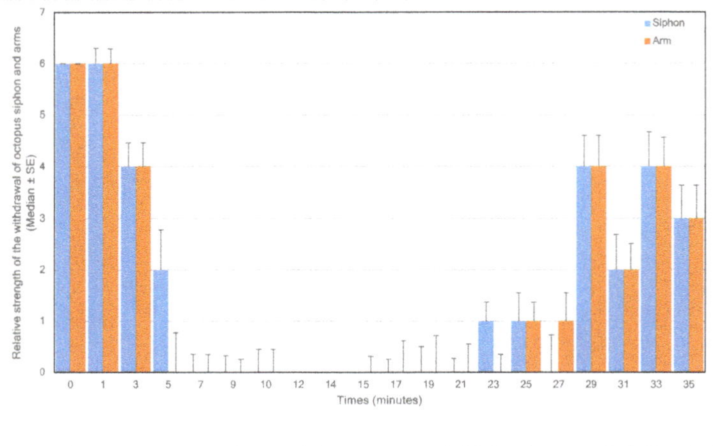

Figure 2. Relative strength of the withdrawal of octopus siphon and arms in response to a touch stimulus versus time (6 = strong, 4 = medium, 2 = low, and 0 = none) in 9 animals; 0–10 min in 1% $MgCl_2$; 10–15 min in 1% $MgCl_2$ and 1% isoflurane; 15–35 min in fresh aerated seawater.

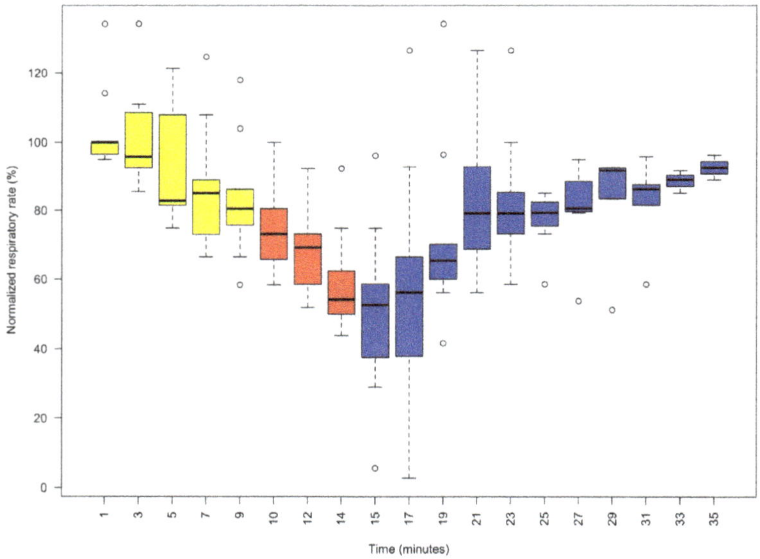

Figure 3. Boxplot of the normalised respiratory rate (%) as determined by the number of mantle contractions per minute in 9 animals. Readings were taken at 2 min intervals. Yellow phase: $MgCl_2$ 1% (10 min). Red phase: isoflurane 1% and 1% $MgCl_2$ (5 min). Blue phase: recovery phase in fresh aerated seawater. As respiratory rate variability between animals is substantial, the outlying circles indicate the recorded maximum and minimum normalised respiratory rates. It should be noted that after 10 min in 1% $MgCl_2$, the respiratory rate stabilised at about 80% of its normalised value, followed by a further gradual decline after 1% isoflurane, which was terminated by changing the medium to clean aerated seawater.

3.2. Addition of 1% Isoflurane

After 10 min in 1% $MgCl_2$, 1% isoflurane was bubbled into the experimental tank. The animals remained unreactive and became paler (Figure 1), but respiration continued at a lower rate as the isoflurane took effect (Figure 3). Reduced respiratory rate is a clear sign that anaesthetics are taking effect in clinical and veterinary settings. After 5 min of anaesthetisation, the isoflurane was switched off, and the normalised respiratory rate was reduced to 54.17% ± 3.59% (median ± se) of the normalised value. At this time, the pale and unresponsive animals were gently moved to a different experimental tank containing fresh aerated seawater to recover.

3.3. Recovery from Anaesthesia

In 6 cases, the animals achieved full recovery of colour pattern, respiratory frequency, reflex arm, and siphon sensitivity in 10–15 min. One animal recovered quickly in 7 min and two others in about 20 min.

Both parameters, colour intensity of the interbrachial membrane and normalised respiratory rate, showed statistically significant differences after 10 min, 15 min and 25 min, compared with data at time 0 (ANOVA one way, $p < 0.05$).

4. Discussion

In the experiments described here, we demonstrated that *Octopus vulgaris* can be pre-anaesthetised with $MgCl_2$ for 10 min and then fully anaesthetised with 1% isoflurane for five minutes with a maximum recovery period of about 20 min.

We observed that the normalised respiratory rate (%) decreased gradually during the first 10 min of the experiment due to the presence of the $MgCl_2$ alone, declining by a normalised value of 20%. Subsequently, adding isoflurane (1%) induced a significant decrease in the respiratory rate, declining by 44%. This effect carried over into the subsequent phase in which the animal was put in fresh aerated seawater but started to decline after a further 2 min, with considerable variation between animals. However, after 20 min (35 min into the experiment overall) of refreshing the bathing solution with clean aerated seawater, all animals recovered, but the normalised respiratory rate did not reach 100% in all animals (median ± standard error, 90.69 ± 1.60). However, all the animals were in good condition. The animals were monitored for the following 7 days, and all were healthy and behaving normally.

This compares very well with our earlier study [20], where increasing doses of isoflurane (0.5 to 2.0% in seawater) were gradually applied to 10 specimens of *Octopus vulgaris* over a period of about 40 min, followed by a recovery period of up to 1 h. Clearly, this new protocol is much less stressful for the animals and utilises much less isoflurane. In addition, the animals showed few signs of discomfort in $MgCl_2$, and there were no flashing responses as the animals became paler in colour and touch reflexes declined as the animals became immobilised. However, in 1% $MgCl_2$ in aerated seawater, the respiratory rate declined by less than 20% and stabilised after about 10 min. Upon the addition of 1% isoflurane, the respiratory rate gradually declined as expected since a declining respiratory indicates an increasing depth of anaesthesia. Furthermore, isoflurane is known to be suitable for maintenance anaesthesia [30] as it also enhances muscle relaxation, while $MgCl_2$ diminishes the early excitatory phase of anaesthesia [31], as we previously suggested [5].

Although magnesium chloride has been used as an anaesthetic for octopods and other cephalopods [16], it is usually applied externally, giving little or no access to the CNS. It is our view that it is wholly inappropriate as an anaesthetic because $MgCl_2$ is a basic muscle relaxant, although it may have minimally sedative properties as mentioned above [20]. We therefore refute the suggestion that it can be used as an anaesthetic substance in its own right as suggested elsewhere [16], where increasing concentrations of $MgCl_2$, up to 3.75%, were used to paralyse specimens of octopus and cuttlefish over about 20 min and were compared with the effects of ethyl alcohol, another non-anaesthetic. Apparently, these substances reversibly depressed evoked activity in the pallial nerve, but no records were shown. In order to clarify nervous activity under these circumstances, recordings of electrical activity need to be clearly demonstrated, perhaps using the recently published methods of Gutnick et al. [21].

Appropriate maintenance anaesthesia for experimentation has not yet been determined for *Octopus vulgaris,* and further experiments are required to determine the most appropriate level of anaesthesia in fresh animals relaxed with $MgCl_2$, which will probably require a slightly higher concentration of isoflurane if the animal is to undergo surgery.

5. Conclusions

1% magnesium chloride is a useful pre-anaesthetic agent for the isoflurane anaesthetisation of *Octopus vulgaris.* It reduces the concentration of isoflurane necessary for complete anaesthetisation and significantly reduces the stress on the animal and the time course of anaesthesia.

Note on the use of vaporisers. The application of volatile general anaesthetics need not involve the use of vaporisers, although they simplify the protocol. Instead, volatile anaesthetics can simply be dissolved in saline at appropriate concentrations, as performed by Dickinson [22]. Isoflurane is usually a controlled substance, not freely available, and is best obtained via collaborating veterinarians.

Author Contributions: Conceptualisation: W.W., A.D.C. and G.P.; Methodology: G.P. and V.M.; Original draft preparation: W.W. and V.M.; Writing and editing: W.W., V.M., A.D.C. and G.P.; Funding acquisition: A.D.C. and G.P.; Formal analysis: V.M., E.C., M.N. and H.K.S.d.Z. All authors have read and agreed to the published version of the manuscript.

Funding: PNRR "MNESYS"—E63C22002170007—PE0000006, University of Naples Federico II.

Institutional Review Board Statement: This research was approved following the European Directive 2010/63 EU L276, the Italian DL. 4/03/2014, n°26, and the ethical principles of Reduction, Refinement, and Replacement (Project n°608/2016-PR-17/06/2016; protocol n°DGSAF 0022292-P-03/10/2017).

Informed Consent Statement: Not applicable.

Data Availability Statement: Data are contained within the article.

Conflicts of Interest: The authors declare no conflict of interest.

References

1. Mather, J.A. Animal suffering: An invertebrate perspective. *J. Appl. Anim. Welf. Sci.* **2001**, *4*, 151–156. [CrossRef]
2. Paolucci, M.; Di Cristo, C.; Di Cosmo, A. Immunological evidence for progesterone and estradiol receptors in the freshwater crayfish Austropotamobius pallipes. *Mol. Reprod. Dev.* **2002**, *63*, 55–62. [CrossRef]
3. Passantino, A.; Elwood, R.W.; Coluccio, P. Why Protect Decapod Crustaceans Used as Models in Biomedical Research and in Ecotoxicology? Ethical and Legislative Considerations. *Animals* **2021**, *11*, 73. [CrossRef]
4. Fiorito, G.; Affuso, A.; Anderson, D.B.; Basil, J.; Bonnaud, L.; Botta, G.; Cole, A.; D'Angelo, L.; De Girolamo, P.; Dennison, N.; et al. Cephalopods in neuroscience: Regulations, research and the 3Rs. *Invert. Neurosci.* **2014**, *14*, 13–36. [CrossRef]
5. Winlow, W.; Polese, G.; Moghadam, H.F.; Ahmed, I.A.; Di Cosmo, A. Sense and Insensibility—An Appraisal of the Effects of Clinical Anesthetics on Gastropod and Cephalopod Molluscs as a Step to Improved Welfare of Cephalopods. *Front. Physiol.* **2018**, *9*, 1147. [CrossRef]
6. Crook, R.J.; Walters, E.T. Nociceptive Behavior and Physiology of Molluscs: Animal Welfare Implications. *Ilar J.* **2011**, *52*, 185–195. [CrossRef] [PubMed]
7. Lewbart, G.A.; Mosley, C. Clinical Anesthesia and Analgesia in Invertebrates. *J. Exot. Pet Med.* **2012**, *21*, 59–70. [CrossRef]
8. Elwood, R.W. Pain and suffering in invertebrates? *Ilar J.* **2011**, *52*, 175–184. [CrossRef] [PubMed]
9. Crook, R.J.; Hanlon, R.T.; Walters, E.T. Squid have nociceptors that display widespread long-term sensitization and spontaneous activity after bodily injury. *J. Neurosci.* **2013**, *33*, 10021–10026. [CrossRef] [PubMed]
10. Sneddon, L.U. Pain in aquatic animals. *J. Exp. Biol.* **2015**, *218*, 967–976. [CrossRef]
11. Crook, R.J. Behavioral and neurophysiological evidence suggests affective pain experience in octopus. *iScience* **2021**, *24*, 102229. [CrossRef]
12. Elwood, R.W. Behavioural Indicators of Pain and Suffering in Arthropods and Might Pain Bite Back? *Animals* **2023**, *13*, 2602. [CrossRef]

13. García-Franco, M. Anaesthetics for the squid Sepioteuthis sepioidea (mollusca: Cephalopoda). *Comp. Biochem. Physiol. Part C Comp. Pharmacol.* **1992**, *103*, 121–123. [CrossRef]
14. Pugliese, C.; Mazza, R.; Andrews, P.L.; Cerra, M.C.; Fiorito, G.; Gattuso, A. Effect of Different Formulations of Magnesium Chloride Used as Anesthetic Agents on the Performance of the Isolated Heart of Octopus vulgaris. *Front. Physiol.* **2016**, *7*, 610. [CrossRef]
15. Butler-Struben, H.M.; Brophy, S.M.; Johnson, N.A.; Crook, R.J. In Vivo Recording of Neural and Behavioral Correlates of Anesthesia Induction, Reversal, and Euthanasia in Cephalopod Molluscs. *Front. Physiol.* **2018**, *9*, 109. [CrossRef]
16. Abbo, L.A.; Himebaugh, N.E.; DeMelo, L.M.; Hanlon, R.T.; Crook, R.J. Anesthetic Efficacy of Magnesium Chloride and Ethyl Alcohol in Temperate Octopus and Cuttlefish Species. *J. Am. Assoc. Lab. Anim. Sci.* **2021**, *60*, 556–567. [CrossRef] [PubMed]
17. James, M.F.M. Clinical Use of Magnesium Infusions in Anesthesia. *Anesth. Analg.* **1992**, *74*, 129–136. [CrossRef] [PubMed]
18. De Oliveira, G.S., Jr.; Castro-Alves, L.J.; Khan, J.H.; McCarthy, R.J. Perioperative systemic magnesium to minimize postoperative pain: A meta-analysis of randomized controlled trials. *Anesthesiology* **2013**, *119*, 178–190. [CrossRef]
19. Fathi Moghadam, H.; Yar, T.; Qazzaz, M.M.; Ahmed, I.A.; Winlow, W. A Comparative Study of Cell Specific Effects of Systemic and Volatile Anesthetics on Identified Motor Neurons and Interneurons of Lymnaea stagnalis (L.), Both in the Isolated Brain and in Single Cell Culture. *Front. Physiol.* **2019**, *10*, 583. [CrossRef] [PubMed]
20. Polese, G.; Winlow, W.; Di Cosmo, A. Dose-dependent effects of the clinical anesthetic isoflurane on *Octopus vulgaris*: A contribution to cephalopod welfare. *J. Aquat. Anim. Health* **2014**, *26*, 285–294. [CrossRef]
21. Gutnick, T.; Neef, A.; Cherninskyi, A.; Ziadi-Künzli, F.; Di Cosmo, A.; Lipp, H.-P.; Kuba, M.J. Recording electrical activity from the brain of behaving octopus. *Curr. Biol.* **2023**, *33*, 1171–1178. [CrossRef] [PubMed]
22. Dickinson, R.; Lieb, W.R.; Franks, N.P. The effects of temperature on the interactions between volatile general anaesthetics and a neuronal nicotinic acetylcholine receptor. *Br. J. Pharmacol.* **1995**, *116*, 2949–2956. [CrossRef]
23. Messenger, J.B. Cephalopod chromatophores: Neurobiology and natural history. *Biol. Rev. Camb. Philos. Soc.* **2001**, *76*, 473–528. [CrossRef] [PubMed]
24. Berge, K.H.; Warner, D.O. Drugs that affect the respiratory system. In *Foundations of Anesthesia, Basic and Clinical Sciences*; Hemmings, H., Hopkins, P., Eds.; Mosby: London, UK, 2000; pp. 491–498.
25. Di Cosmo, A.; Polese, G.; Bertapelle, C.; Palumbo, A.L.; Zullo, L.C. *Cefalopodi. Benessere ed Animal Care Dell'animale da Laboratorio*; Le Point Veterinaire Italie: Milano, Italy, 2015.
26. Maselli, V.; Polese, G.; Al-Soudy, A.S.; Buglione, M.; Di Cosmo, A. Cognitive Stimulation Induces Differential Gene Expression in *Octopus vulgaris*: The Key Role of Protocadherins. *Biology* **2020**, *9*, 196. [CrossRef]
27. Maselli, V.; Al-Soudy, A.S.; Buglione, M.; Aria, M.; Polese, G.; Di Cosmo, A. Sensorial Hierarchy in *Octopus vulgaris*'s Food Choice: Chemical vs. Visual. *Animals* **2020**, *10*, 457. [CrossRef]
28. Collins, T.J. ImageJ for microscopy. *Biotechniques* **2007**, *43*, 25–30. [CrossRef]
29. Team, R.C. *R: A Language and Environment for Statistical Computing*; R Foundation for Statistical Computing: Vienna, Austria, 2018.
30. Pauca, A.L.; Dripps, R.D. Clinical experience with isoflurane (Forane): Preliminary communication. *BJA Br. J. Anaesth.* **1973**, *45*, 697–703. [CrossRef]
31. Guedel, A.E. Inhalation Anesthesia: A Fundamental Guide. *Anesth. Analg.* **1937**, *16*, 119–120. [CrossRef]

animals

Article

Invisible Invertebrates: The Welfare of Invertebrates in Public Aquaria

Kerry Perkins

Aquaculture, Fisheries and Marine Studies, University Centre Sparsholt, Winchester SO21 2NF, UK; kerry.perkins@sparsholt.ac.uk

Simple Summary: Aquatic invertebrates are understudied and often overlooked in public aquarium welfare. By determining public opinion on aquatic invertebrate welfare, we can obtain a better picture on whether they are even considered during an aquarium visit. Using TripAdvisor, it was determined that aquatic invertebrates were under-represented in negative welfare comments compared with their vertebrate counterparts. The public were more likely to comment on welfare issues if from the UK or USA. To aid in a better understanding of aquatic welfare by the public, zoo and aquarium associations should help create guidelines for aquatic invertebrates. Guideline creation would help increase the construction of welfare indicators and support welfare tool kits. This would hopefully result in the better consideration of invertebrate welfare and public understanding in public aquaria.

Abstract: Awareness of welfare issues within animal collections is increasing as information becomes more accessible for staff and the public. A knowledge gap remains when considering the welfare of invertebrates, particularly when housed in public aquaria. TripAdvisor comments were analyzed for 485 worldwide aquariums. The public focused on anthropogenic features or charismatic organisms within collections. Invertebrate welfare was only presented in 18% of negative welfare comments compared with the 51% of represented vertebrates and 31% of negative general welfare comments. The UK and USA reported a greater number of perceived invertebrate welfare issues. Greater dissemination of information between aquarists and scientists should be encouraged to drive welfare standards and improve husbandry. In addition, incorporating input from invertebrate aquarists while utilizing welfare toolkits are vital for improving overall standards if we are to have greater representation of invertebrate welfare in public aquaria.

Keywords: invertebrate; aquarium; welfare

1. Introduction

The definition of animal welfare has been considerably debated within the literature and is often adjusted depending on the lens of the observer [1–3]. Within zoos and aquariums, welfare is often based around Mellor's five domains model [4] which has five domains split into nutrition, environment, health, behavior and mental state. Both the World Association of Zoos and Aquariums (WAZA) and the European Association of Zoos and Aquaria (EAZA) [5] have adopted Mellor's model to monitor animal welfare status within collections. What is considered good animal welfare in a captive setting maybe different from wild animals due to different pressures acting on the animal [3]. Subsequently, priorities in an animal's life may be altered if it is in a captive setting. Aspects such as life cycle, avoidance of predators and access to food may be considerably different compared to their wild counterparts. Therefore, the careful creation of environments within aquaria are important, as a large number of animals displayed are still captured from the wild [6].

Understanding of wild animal welfare and human input is still debated [1–3,7], particularly in species with limited knowledge of conditions in the wild and their life cycles. A captive environment may also contain different goals and husbandry settings than that of the wild. Within the UK, laboratory settings base animal numbers and welfare on the 3R model of replacement, reduction, and refinement, by either replacing the animal use with alternatives such as simulations, reducing the number of animals used or refining the experiment methodology. This reduces the overuse of animals within scientific research. Legislation also gives additional protection, welfare conditions and guidance on vertebrates within these situations [8,9]. Similarly, within zoo and aquarium populations, vertebrates are also given welfare considerations, often with a different focus, due to the conservation and breeding focus of many zoos [10]. In contrast, invertebrates do not necessarily receive the same consideration both in laboratory and zoo collections. Only some invertebrate groups such as cephalopods and decapods are protected in certain countries such as New Zealand [11,12]. This lack of consideration for invertebrates is often not due to lack of addressing the needs, but rather the lack of understanding that they have needs.

With an estimated 97% of organisms being invertebrates, the overall understanding and research into their welfare despite their numbers is not comparable to that of vertebrates [13]. This is particularly evident when concerned with welfare indicators and positive/negative responses to stimuli. Debate on the pain perception of vertebrates and invertebrates has contributed to discussion on whether there is a capacity to suffer within certain species [14]. Recent studies in higher invertebrates such as cephalopods and crustaceans have shown the capacity for nociception [15,16]. Our understanding of pain experienced by invertebrates, however, is still limited [17]. This, therefore, raises the question around the potential capacity to suffer by invertebrates. The absence or difference of comparative organs such as a nervous system does not necessarily indicate a lack of negative or painful experiences for the organism [16]. Therefore, the ability to feel pain should not be the sole determining factor of poor welfare, particularly when using welfare models, and we should be cautious of using this metric alone.

Many animal welfare models are based off the requirements of terrestrial vertebrates. The five freedom model was developed from a 1965 UK government report on livestock husbandry with formalization in 1979 [18]. This key piece of guidance has given a base from which other models have been constructed. However, some aspects of the five freedoms such as 'freedom from thirst' have little relevance for aquatic species. Within the aquarium and zoo community, Mellor's five domain model is the most popular [1]. Although useful for terrestrial vertebrates, its suitability for other species is still yet to be fully investigated [18,19]. Another criticism is that, like similar models, it is often based around a single point in time rather than the lifetime of experiences of an animal [20]. As a result, welfare tools based off the five domains have been constructed for utilization within zoos and aquaria instead. The Animal Welfare Assessment Grid (AWAG) has been tested in a variety of zoo and aquarium situations [20–22]. The AWAG, however, has not been trialed extensively within multispecies exhibits or with aquatic colonial species.

Olge and DeSmet [7] suggest one way of improving welfare may be the inclusion of zookeepers in monitoring and assessment, and to use a more contemporary approach with a subjective measure of behavior and animal personality. There is, however, a question around whether zookeepers have sufficient understanding of welfare and whether it is a priority within their collection. It was suggested that by increasing staff utilization in the assessment, they may proactively meet the needs of the individual animal through the input of the animal staff. It is unclear, however, whether this would translate into an aquarium situation due to multiple species exhibits or potential bias arising towards vertebrates, such as prioritizing fish and mammals. It is, therefore, important to investigate the available resources to improve aquatic invertebrates and consider the current public perception of their welfare.

A call to action by Oldfield and Bonano [22] highlighted the urgent need for zoos and aquaria to conduct behavior studies on bony fishes due to their under representation within the literature. Similarly, it is not through a lack of aquatic invertebrates within the collections, but the lack of study, which is reflected in the underrepresentation. Our custodial responsibilities to the animals are constantly improving and advancing; however, our improvements in welfare are still questionable [23]. In aquariums, this has particularly been evident with the increase in public interest documentaries such as Blackfish [24] and The Cove [25]. Historically, aquatic mammals were considered ambassadors for their species; however, due to recent documentaries, there is now a question around the validity of this claim [26]. With an increased awareness of and interest in welfare in aquariums, we should be encouraging better transparency of our welfare indicators. Better transparency may increase public confidence with collections, but also to allow the better understanding of behaviors that the organism may be exhibiting. One way to potentially standardize welfare monitoring is through tools provided by regional or global zoos and aquarium accreditation schemes.

Currently, the public awareness of zoo association programs is under-researched [27]. The understanding of aquarium welfare indicators by the public is also potentially limited, with most studies focusing on zoos [28]. Studies within zoo species have shown that the public often view certain behaviors as a lower standard of welfare, such as repetitive movement rather than resting [28,29]. Melfi et al. [30] also found that most visitors believe they understand animal welfare intuitively. When attending aquariums, the public's motivation will vary and their preconceived ideas will be influenced by their existing knowledge and experience [22]. Prior experience combined with emotion can play an important role, with visitors experiencing a multi-layered zoo interaction as having a more positive perception [31]. Similar comments could be made about aquarists, who care for aquatic invertebrates, about how their existing history may influence overall day-to-day welfare decisions.

It is, therefore, important that we identify areas of potential influence and bias by the members of the public and potentially the aquariums themselves, if we are to fully understand welfare decisions. The rise of social media has given a platform for people to share their thoughts and views more freely [32]. Review sites have gained popularity as the world becomes more connected, giving people the opportunity to share their experiences [33]. By reviewing tourism businesses, the customers experience can be communicated to prospective patrons. Because of the ability to contribute to other people's experience, websites such as Yelp and TripAdvisor have become a popular resource. Since its launch in 2000, TripAdvisor has acted as a guide to the public, but from 2004 user-generated content has been the main focus of the site, reaching 100's of millions of visitors a month [34]. The popularity of the site has gained interest from researchers on sentiment analysis, complaints and satisfaction [32–34]. Because of the multiple aspects such as ethic, welfare and care comments, wildlife and tourism researchers have used TripAdvisor to understand the impact of wildlife encounters and drivers to ecotourism [35–38]. It is this investigation into welfare and public opinion of animal attractions that can aid us in understanding and potentially improving welfare within aquariums. In this article, we will investigate how the TripAdvisor platform may be used to gain insight into the public's perception of welfare in aquarium. In addition, we will consider the potential actions collections can implement to assist in increasing invertebrate welfare knowledge of visitors.

2. Materials and Methods

A list of all aquariums on TripAdvisor was compiled in September 2020. This was conducted by utilizing the attractions function on TripAdvisor, which has a category for zoos and aquariums. In addition the search term, aquarium was inputted into the search function of the website to discover any additional aquariums that may have been missed due to not being included in the attractions category. Aquariums within Zoos and theme parks were excluded from analysis. This was due to the fact that keywords and comments around the aquarium could not be separated from other comments about the attraction. The location and number of reviews were recorded along with the rating of the aquarium in respect to the trip advisor scale of 1–5, (Terrible = 1, Poor = 2, Average = 3, Very good = 4, and Excellent = 5). Keywords of 'Welfare', 'Dead', 'Sad', 'Horrible', 'Confinement' and 'Torture' were then used to search the text for possible negative welfare implications.

Keywords were selected from terms used by Walker et al. [39] as well as other terms that had been used by animal rights groups to explain aquarium welfare [40]. Only negative welfare comments were included due to the ambiguous and often unclear positive welfare comments within the text. Keywords were also translated into the recognized language of the aquarium's country via google translate to reduce loss of data, this equated to 31 languages in total. The number of occurrences of the keyword were then recorded. Where the word was not used in respect to animal welfare, it was excluded from the data set. Results were broken down into major geographical regions determined by the location of the aquarium. In a large number of cases, the native language was the predominant number of reviews, suggesting that visitors were likely to be domestic to the country rather than international tourists.

Overall, comments were also analyzed around negative perceptions such as dirty tanks or tanks too small for the animal; the data set was then further refined to just include aquatic invertebrates for further analysis. The comments were divided by invertebrate species to identify differences. This also allowed for identifying other terms, such as sick, damaged and injured, that were not picked up in the key terms search specifically relating to invertebrates. ANOVAs were performed on the data in Minitab v20 between regions with Bonferroni post hoc tests to identify specific differences.

3. Results

3.1. Overall Visitor Satisfaction

A total of 485 aquariums were identified on TripAdvisor from 77 countries. The average rating for all aquariums was 3.95 out of 5, confirming a very good experience overall. However, there was variation between geographic regions on the overall experience (Table 1). North American visitors were more likely to rate their experience as excellent ($p < 0.001$). European aquariums, however, had a greater number of terrible ratings than other regions ($p < 0.05$). Higher ratings were given to aquariums with a greater number of comments, with the exception of excellent. The rating of excellent was close to the mean (1489) of comments rather than at the maximum, suggesting that an excellent experience was not necessarily due to the number of comments provided.

Table 1. Overall customer satisfaction ratings divided by geographic regions. Values are given in percentages for the overall ratings (Terrible, Poor, Average, Very good, Excellent).

Region	Terrible	Poor	Average	Very Good	Excellent
Africa	4.7	4.1	19.8	36.5	34.9
Asia	3.7	5.3	19.0	35.0	37.0
Europe	6.3	9.4	20.2	28.8	35.3
North America	2.1	3.3	12.5	30.7	51.4
Oceania	2.1	5.4	18.1	36.3	38.1
South America	4.5	3.8	18.9	34.6	38.1

3.2. Visitor Views on Welfare

3.2.1. Keywords

Sad was the most prevalent keyword, appearing 2605 times out of 3984 key words found within the comments. The least-used keyword was confinement (37), followed by torture (72), dead (375), welfare (416) and horrible (479). The word sad was often used as a comment on the status or mood of an animal; for example, penguins were often described as sad, along with dolphins, seals and turtles. The regions that had the greatest number of comments were Asia, Europe and North America. Although having similar number of aquariums within each region, there was a significant difference between the regions with the term sad and other terms that were used ($p < 0.001$) (Figure 1).

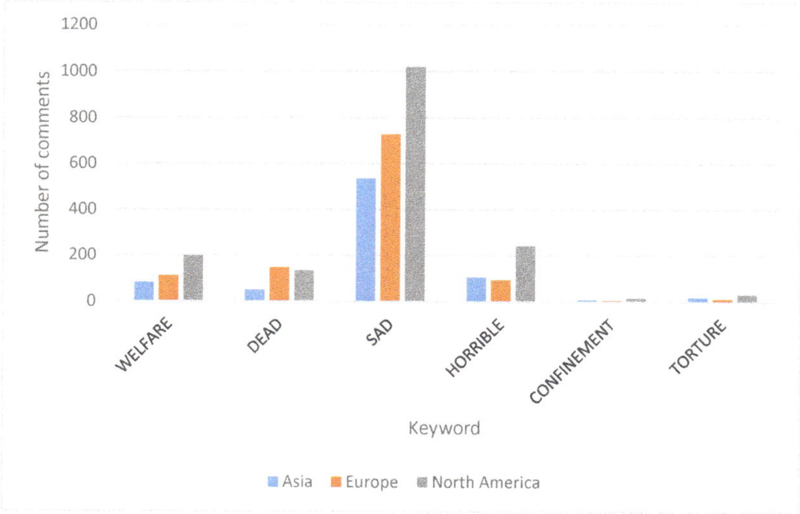

Figure 1. Keywords used within the comments of the three largest regions. The keyword sad had the greatest usage when describing welfare in all three regions. Confinement and torture were not heavily utilized within the comments.

3.2.2. Negative Comments

All aquariums had comments from their visitors on TripAdvisor, ranging from their visitor experience, animal experiences and welfare. When just focusing on negative welfare comments, 328 aquariums had at least one comment. Out of the 77 countries, 42% of the countries had at least one negative welfare comment for every aquarium within their country. In contrast, only 10% had no negative welfare comments for the aquariums within their country. However, these were generally represented by countries with just one aquarium present within their borders. The North American region had the highest level of negative comments per number of aquariums (75%) whereas Asia had the fewest (64%). Comments were often related to water quality, tank size or sick fish.

3.2.3. Aquatic Invertebrate Comments

Out of 77 countries, 29 had comments of negative welfare concerning aquatic invertebrates. The number of specific comments, however, were low in number (18%) compared to that of vertebrates (51%) and general welfare (31%) There were no aquatic invertebrate comments within the South American region and limited numbers (n = 2) within the African region and Oceania region (n = 6). The highest number of comments was in the Europe region (n = 43), with the largest number coming from the UK (n = 24). North America also showed a high number of comments (n = 37), with the largest number coming from a sole country, the USA (n = 29).

When divided into individual species, octopus was mentioned significantly ($p < 0.01$) more than other species (n = 43) (Figure 2). Several comments were based around tank size and enrichment of the environment. Invertebrate touch pools were the next highest (n = 23), with comments based around excessive force and lifting animals out of the water. Jellyfish had a greater number of comments about being damaged or dead compared to the rest of the invertebrate welfare comments.

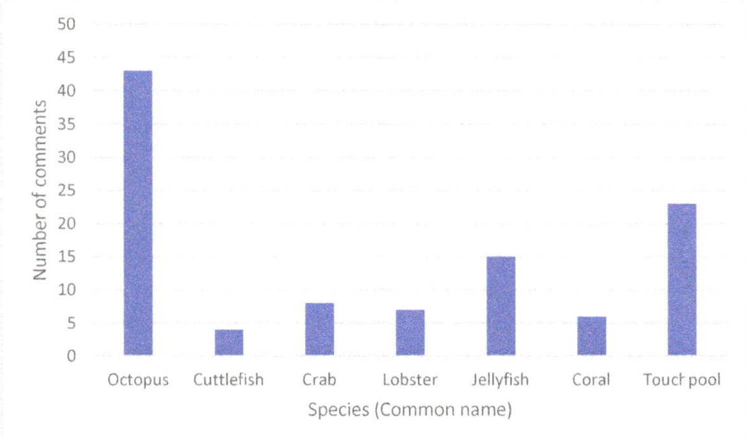

Figure 2. Number of negative welfare comments divided by species; note that touch pools included were containing only invertebrates, with most comments around starfish handling.

The European region had welfare comments in all invertebrate groups. Similarly, the North American region showed similar numbers to Europe except in cuttlefish ($n = 0$) and crab ($n = 2$), where levels of comments were lower. The only regions to have negative comments in respect to crabs were Europe and North America. Asia showed the greatest number of comments in relation to coral welfare ($n = 4$) compared to other regions.

4. Discussion

4.1. Visitor Perception

By investigating key terms and comments within TripAdvisor, we started to gain insight into the public view of negative aquarium welfare. Emotive words such as sad were used more frequently in comments than the word welfare. The term sad was frequently used in conjunction with particular animals, such as penguins and dolphins. This is possibly due to preconceived ideas around the animals emotional state and vertebrate bias [38]. Species which people connect with on a visit are more likely to incite conservation action for that individual species [41]. It is possible the public may attend to or be specifically interested in certain animals, and therefore are more likely to mention welfare concerns in regard to those animals. Both Europe and North America had a greater number of keywords overall, despite Asia having a similar number of aquariums, suggesting cultural and societal ideology may play a role in how the public perceive welfare. When surveying Chinese and European zookeepers, Bacon et al. [1] found that there was variation in welfare perception; it is possible a similar variation exists within the general public. This may also be reflected in the aquatic invertebrate comments, with corals having a greater representation in Asia than other regions due to their commercial importance for the region.

The overall lower number of poor welfare comments of aquatic invertebrates could be due to numerous factors. Comments concerning vertebrates were emotive, reflecting trends shown by the keywords. It may be that aquatic invertebrates do not elicit the same emotional response in the public. However, if it was lack of emotional involvement, we should see similar numbers within each species, or possibly no comments at all. Instead, we found that octopus received the most comments, with most focused around tank size. This is an interesting observation, as most octopus species have not had their required tank size scientifically investigated. The Giant Pacific Octopus [GPO} Association of Zoos and Aquariums (AZA) tank size is based off aquarists' experience. Holmberg [42] criticized the lack of scientific rigor when setting the GPO tank size by not utilizing the peer-reviewed literature and only references from aquarists and surveys. By solely using this approach, it does not necessarily account for the animal's overall need for space, nor does it use the wild populations' behavior to help inform a tank suitable to exhibit a range of natural behaviors.

Another possibility for the low number of aquatic invertebrate welfare comments may be due to the general understanding of natural behaviors of the animal. This would be particularly evident in species that the public may not have encountered before, and could possibly explain the lower numbers in species such as cuttlefish and coral. Despite showing cognitive ability such as self-control [43] and observational learning, cuttlefish [44] do not receive as much recognition as octopus. If the public were to possibly understand more about cuttlefish cognitive abilities, would they receive similar numbers of comments? It is most likely that a combination of a lack of understanding of welfare cues within the species combined with a familiarity bias has contributed to underrepresentation of comments within this species.

Overall, we could question the reliability of members of the public in identifying welfare concerns, as they are not trained ethologists or even aquarists. Comments including touch pool areas seemed to indicate some understanding of the basic needs of the animals. Comments were around a lack of care from other members of the public or insufficient monitoring by staff. As touch pools are often points of contact with animals, having poor supervision gives a greater potential for animals to be harmed. The harming of animals may be easier for the public to determine even without specialist knowledge of the species. Activities such as lifting animals out of the water or excessive force when touching the animals would be easily identifiable to other members of the public. This may give them more confidence in reporting potential breaches of welfare.

What is potentially interesting is that even if there were comments of poor welfare, it did not necessarily reflect in a lower rating in the aquarium. In some cases, ratings were still 4 out of 5 stars even when commenting on the welfare of an individual organism. People are often drawn to the possibility of having direct and indirect physical contact with a wild animal, which can increase collection appeal [45]. Whether people are willing to compromise on welfare or are aware of the possible implications of direct contact has not been thoroughly investigated. In the North American region, the number of excellent ratings was significantly higher than that of other regions. Despite such high ratings, it was also one of the regions with the highest number of negative welfare comments in relation to aquatic invertebrates and overall.

It is possible that by using TripAdvisor, there may be a country or regional bias introduced. TripAdvisor is an American website and, therefore, the American public may have more familiarity or desire to use it. Analysis of utilization has shown that popularity is actually more global, with European, Oceania, South America and China utilizing the website [33], suggesting that although it may be a factor in commenting, it is less likely to be the sole reason for its use. The UK also showed a greater number of poor welfare comments; however, unlike the USA, they showed the greatest number of terrible ratings of any region. Vasquez [46] suggested that people are more likely to complain online about failure to meet expectations or disappointment than in person, most likely due to the anonymity that they receive. This would possibly explain why welfare comments would be present on a website that is effectively reviewing experiences rather than reporting concerns directly to the aquarium. It is, therefore, suggested that as an aquarium, reviewing negative comments on TripAdvisor would help gain insight into the visitors' perception of your animal welfare. It would also identify areas where further education is needed for the public to understand welfare indicators; however, to truly achieve this goal, further studies are needed, particularly concerning invertebrates.

4.2. Aquarium Welfare Resources

Husbandry guidelines can be invaluable as a starting point to form procedures and welfare indicators. When the number of guidelines is compared between aquatic invertebrates and vertebrates, there is a considerable difference (Table 2). Aquatic invertebrates are underrepresented in all three of the associations examined. In AZA guidelines, aquatic invertebrates and aquatic vertebrates have similar numbers, which are considerably lower than the terrestrial guidelines. The British and Irish Association of Zoos and Aquariums (BIAZA) has a better representation of aquatic invertebrates; however, this may be due to the guidelines being shorter in length and therefore easier to compile. Within the Europe Association of Zoos and Aquariums (EAZA), there are no guidelines available for aquatic invertebrates. Even with terrestrial invertebrates, there are ten times more vertebrate guidelines, even though there are numerous invertebrates within the zoo and aquarium collections. It is, therefore, important that we put greater importance on compiling aquatic invertebrate guidelines. A potential starting point could be to determine welfare indicators while we build husbandry guidelines.

Table 2. Zoo associations and the number of produced guidelines as of February 2023.

Association	Terrestrial Vertebrates	Terrestrial Invertebrates	Aquatic Vertebrates	Aquatic Invertebrates
AZA	28	0	3	3
BIAZA	44	17	12	7
EAZA	50	5	5	0
Total	132	22	20	10

Staff time is often at a premium within aquariums and, therefore, easy-to-implement welfare data collection is vital to success. Understanding of the welfare model and how data collection will be input into welfare decisions may also aid in motivating their collection. Kagan et al. [23] established that conservation and welfare require different priorities, task and goals. It requires almost a mind shift to see how everyday tasks can be incorporated to help advance welfare. By creating a positive environment, the ability to understand the animal better would result in overall standards improving [47]. The human factor of good animal welfare in collections was summarized by Bacon et al. [1], and one of the key findings was that the aquarium should consider the wellbeing of their staff as a function of welfare. Aquarists have a large influence on the organisms that they care for, and if their welfare is compromised then potentially so are the animals. It is important that association bodies also assist in providing guidance and support for data collection. Co-ordination by the associations could also help facilitate collaboration between scientific researchers and aquarists at aquariums.

As our understanding and technology improves, so does the understanding of the animals within our care. Despite this, many aquatic invertebrates remain under investigated, particularly in our understanding of effective states. It is suggested that cognitive bias measurements are adaptable for all species due to the benefit of identifying rewards versus danger [48]. Without the basic understanding of effective states in numerous invertebrate species, being able to use cognitive bias for welfare measurements may be difficult. Using scientific methods such as Go/No-Go tasks and using spatial cues may help start determining bias effects within the lesser-studied aquatic invertebrates [49]. A more scientific approach may also suit the capabilities of those caring for the animals, as it is reported that 88% of USA aquarists have a bachelor's degree or above [50]. This high level of education would suggest the capacity to learn, collect data, implement it and contribute to invertebrate welfare.

Welfare tool kits could possibly provide standardization and would allow for aquarists to use pre-existing knowledge of the animal. One of the most utilized tool kits within zoos is the Animal Welfare Assessment Grid (AWAG). Its popularity is due to its versatility with different species. The AWAG also allows for a temporal component that many other welfare measures are lacking. A component of research is required for setting up the appropriate indicators and can involve veterinarians, aquarists and the literature. Animal individuality and its possible change over its lifetime gives a versatility that many other models are lacking. In respect to aquatic invertebrates, the AWAG has unfortunately had limited implementation. Narshi et al. [17] successfully demonstrated the use of the AWAG on cephalopods and decapods, although they suggested further study and utilization is needed to test its full potential for aquatic invertebrates. By investigating aquatic invertebrate species utilizing the AWAG, data can be generated and tested over multiple collections. In addition, the sharing of data and indicators is something that zoo and aquarium associations could co-ordinate. It is this collaborative aspect within the aquarium community that will help increase our understanding of aquatic invertebrate welfare.

5. Conclusions

A knowledge gap still exists around aquatic invertebrate welfare in public aquaria. Awareness of negative welfare indicators by the public is limited to a few species and often emotive when being discussed. TripAdvisor may offer an insight into the public's overall understanding and opinions of welfare within an aquarium due to the anonymity. Europe and North American regions have a greater number of comments in respect to aquatic invertebrate welfare compared to other regions.

Aquarium and zoo associations should be encouraged to help facilitate and produce guidelines for more aquatic invertebrate species. By making use of the scientific and veterinary communities, welfare indicators in welfare tool kits can become more robust. In addition, incorporating input from invertebrate aquarists are vital for improving overall standards if we are to have greater representation of invertebrate welfare in public aquaria. As such, aquarium collections need to support and train their staff in welfare tools and consider their role in championing welfare.

Funding: This research received no external funding.

Institutional Review Board Statement: The study was conducted in accordance with the Declaration of Helsinki, and approved by the Institutional Review Board (or Ethics Committee) of University Centre Sparsholt (protocol code UCSEC_1122 on 22 October 2019).

Informed Consent Statement: Not applicable.

Data Availability Statement: The data presented in this study are available on request from the corresponding author.

Conflicts of Interest: The author declares no conflict of interest.

References

1. Bacon, H.; Bell, C.; Dwyer, C.M.; Waran, N.; Qing, Y.; Xia, L.; Shaw, D.J. Exploration of cultural norms and behavioural beliefs about zoo animal behaviour, welfare, ethics and husbandry practices in a sample of the international zoo community. *Zoo Biol.* **2023**, *42*, 416–428. [CrossRef] [PubMed]
2. Cole, J.; Fraser, D. Zoo Animal Welfare: The Human Dimension. *J. Appl. Anim. Welf. Sci.* **2018**, *21* (Suppl. S1), 49–58. [CrossRef] [PubMed]
3. Veasey, J.S. Differing animal welfare conceptions and what they mean for the future of zoos and aquariums, insights from an animal welfare audit. *Zoo Biol.* **2022**, *41*, 292–307. [CrossRef] [PubMed]
4. Mellor, D.J. Operational details of the five domains model and its key applications to the assessment and management of animal welfare. *Animals* **2017**, *7*, 60. [CrossRef] [PubMed]
5. EAZA. Available online: www.eaza.net (accessed on 28 August 2023).
6. Id, B.F.; Pempek, J.; George, K.A.; Flint, J.; Id, T.W.; Flint, M. Using ecosystem health and welfare assessments to determine impacts of wild collection for public aquariums. *PLoS ONE* **2023**, *18*, e0285198.
7. Ogle, B.; DeSmet, A. The perception of felid welfare by zookeepers in North America and the implications for zoo managers. *Zoo Biol.* **2023**, *42*, 651–660. [CrossRef] [PubMed]
8. Hobson-West, P. The role of 'public opinion' in the UK animal research debate. *J. Med. Ethics* **2010**, *36*, 46–49. [CrossRef]
9. Davies, G.; Greenhough, B.; Hobson-West, P.; Kirk, R.G. Science, culture, and care in laboratory animal research: Interdisciplinary perspectives on the history and future of the 3Rs. *Sci. Technol. Hum. Values* **2018**, *43*, 603–621. [CrossRef]
10. Binding, S.; Farmer, H.; Krusin, L.; Cronin, K. Status of animal welfare research in zoos and aquariums: Where are we, where to next? *J. Zoo Aquar. Res.* **2020**, *8*, 166–174.
11. Pollo, S.; Vitale, A. Invertebrates and humans: Science, ethics, and policy. In *The Welfare of Invertebrate Animals*; Carere, C., Mather, J., Eds.; Springer: Cham, Switzerland, 2019; pp. 7–22.

12. Morris, R.K.A.; Welch, M.D. Is invertebrate conservation in Great Britain best achieved by policies that increase species protection? *J. Insect Conserv.* **2023**, *27*, 527–531. [CrossRef]
13. Mather, J.A. Animal Suffering: An Invertebrate Perspective. *J. Appl. Anim. Welf. Sci.* **2001**, *4*, 151–156. [CrossRef]
14. Crook, R.J. The welfare of invertebrate animals in research: Can science's next generation improve their lot? *Postdoc J.* **2013**, *1*, 9–20. [CrossRef]
15. Elwood, R.W. Hermit crabs, shells, and sentience. *Anim. Cogn.* **2022**, *25*, 1241–1257. [CrossRef] [PubMed]
16. Sneddon, L.U. Comparative physiology of nociception and pain. *Physiology* **2018**, *33*, 63–73. [CrossRef] [PubMed]
17. Narshi, T.M.; Free, D.; Justice, W.S.M.; Smith, S.J.; Wolfensohn, S. Welfare Assessment of Invertebrates: Adapting the Animal Welfare Assessment Grid (AWAG) for Zoo Decapods and Cephalopods. *Animals* **2022**, *12*, 1675. [CrossRef]
18. Ryan, M.; Waters, R.; Wolfensohn, S. Assessment of the welfare of experimental cattle and pigs using the animal welfare assessment grid. *Animals* **2021**, *11*, 999. [CrossRef] [PubMed]
19. Perkins, K. Can Aquatic Invertebrates within Public Aquaria Fit the Five Domain Welfare Model? *J. Appl. Anim. Ethics Res.* **2021**, *3*, 181–204. [CrossRef]
20. Ma, S.A.; Kang, H.J.; Lee, K.; Kim, S.A.; Han, J.S. Animal Welfare Assessment in 16 Zoos in South Korea Using the Modified Animal Welfare Assessment Grid. *Front. Vet. Sci.* **2022**, *9*, 860741. [CrossRef]
21. Wolfensohn, S.; Shotton, J.; Bowley, H.; Davies, S.; Thompson, S.; Justice, W.S.M. Assessment of welfare in zoo animals: Towards optimum quality of life. *Animals* **2018**, *8*, 110. [CrossRef]
22. Oldfield, R.G.; Bonano, P.E. Psychological and social well-being of bony fishes in zoos and aquariums. *Zoo Biol.* **2022**, *42*, 185–193. [CrossRef]
23. Kagan, R.; Allard, S.; Carter, S. What Is the Future for Zoos and Aquariums? *J. Appl. Anim. Welf. Sci.* **2018**, *21* (Suppl. S1), 59–70. [CrossRef] [PubMed]
24. Cowperthwaite, G. *Blackfish*; Magnolia Pictures: New York, NY, USA, 2013.
25. Psihoyos, L. *The Cove*; Lionsgate: Santa Monica, CA, USA, 2009.
26. Boissat, L.; Thomas-Walters, L.; Veríssimo, D. Nature documentaries as catalysts for change Mapping out the 'Blackfish Effect'. *People Nat.* **2021**, *3*, 1179–1192. [CrossRef]
27. Warsaw, D.; Sayers, J. The influence of animal welfare accreditation programmes on zoo visitor perceptions of the welfare of zoo animals. *J. Zoo Aquar. Res.* **2020**, *8*, 188–193. [CrossRef]
28. Miller, L.J.; Luebke, J.F.; Matiasek, J. Viewing African and Asian elephants at accredited zoological institutions: Conservation intent and perceptions of animal welfare. *Zoo Biol.* **2018**, *37*, 466–477. [CrossRef]
29. Miller, L.J. Visitor reaction to pacing behavior: Influence on the perception of animal care and interest in supporting zoological institutions. *Zoo Biol.* **2012**, *31*, 242–248. [CrossRef] [PubMed]
30. Melfi, V.A.; McCormick, W.; Gibbs, A. A preliminary assessment of how zoo visitors evaluate animal welfare according to enclosure style and the expression of behavior. *Anthrozoos* **2004**, *17*, 98–108. [CrossRef]
31. Godinez, A.M.; Fernandez, E.J. What is the zoo experience? How zoos impact a visitor's behaviors, perceptions, and conservation efforts. *Front. Psychol.* **2019**, *10*, 1746. [CrossRef] [PubMed]
32. Ahmed, Y.A.; Ahmad, M.N.; Ahmad, N.; Zakaria, N.H. Social media for knowledge-sharing: A systematic literature review. *Telemat. Inform.* **2019**, *37*, 72–112. [CrossRef]
33. Limberger, P.F.; Dos Anjos, F.A.; de Souza Meira, J.V.; dos Anjos, S.J.G. Satisfaction in hospitality on TripAdvisor.com: An analysis of the correlation between evaluation criteria and overall satisfaction. *Tour. Manag. Stud.* **2014**, *10*, 59–65.
34. Cambria, E.; Valdivia, A.; Luzón, M.V.; Herrera, F. Affective Computing and Sentiment Analysis Sentiment Analysis in TripAdvisor. 2017. Available online: www.computer.org/intelligent (accessed on 12 July 2023).
35. Jones, P.; Comfort, D. Tourism Companies and Animal Welfare. *Athens J. Tour.* **2021**, *8*, 75–88. [CrossRef]
36. Moorhouse, T.P.; Dahlsjö, C.A.L.; Baker, S.E.; D'Cruze, N.C.; Macdonald, D.W. The customer isn't always right—Conservation and animal welfare implications of the increasing demand for wildlife tourism. *PLoS ONE* **2015**, *10*, e0138939. [CrossRef] [PubMed]
37. Ziegler, J.A.; Silberg, J.N.; Araujo, G.; Labaja, J.; Ponzo, A.; Rollins, R.; Dearden, P. A guilty pleasure: Tourist perspectives on the ethics of feeding whale sharks in Oslob, Philippines. *Tour. Manag.* **2018**, *68*, 264–274. [CrossRef]
38. Powell, D.M.; Bullock, E.V.W. Evaluation of factors affecting emotional responses in zoo visitors and the impact of emotion on conservation mindedness. *Anthrozoos* **2014**, *27*, 389–405. [CrossRef]
39. Walker, M.; Diez-Leon, M.; Mason, G. Animal welfare science: Recent publication trends and future research priorities. *Int. J. Comp. Psychol.* **2014**, *27*, 80–100. [CrossRef]
40. Casamitjana, J. *Aquatic Zoos. A Critical Study of UK Public Aquaria in the Year 2004*; The Captive Animals' Protection Society: Manchester, UK, 2004.
41. Skibins, J.C.; Powell, R.B.; Hallo, J.C. Charisma and conservation: Charismatic megafauna's influence on safari and zoo tourists' pro-conservation behaviors. *Biodivers. Conserv.* **2013**, *22*, 959–982. [CrossRef]
42. Holmberg, M. Crisis Ecology at the Vancouver Aquarium: Putting Octopuses to Work for Conservation. Ph.D. Dissertation, University of British Columbia, Vancouver, BC, Canada, 2019.
43. Schnell, A.K.; Boeckle, M.; Rivera, M.; Clayton, N.S.; Hanlon, R.T. Cuttlefish exert self-control in a delay of gratification task. *Proc. R. Soc. B Biol. Sci.* **2021**, *288*, 20203161. [CrossRef]
44. Huang, K.L.; Chiao, C.C. Can cuttlefish learn by observing others? *Anim. Cogn.* **2013**, *16*, 313–320. [CrossRef]

45. Cruze, N.D.; Khan, S.; Carder, G.; Megson, D.; Coulthard, E.; Norrey, J.; Groves, G. Modern Zoos and Aquariums and Their Implications for Wild Animal Welfare. *Animals* **2019**, *4*, 332.
46. Vásquez, C. Complaints online: The case of TripAdvisor. *J. Pragmat.* **2011**, *43*, 1707–1717. [CrossRef]
47. Burton, R.J.F.; Peoples, S.; Cooper, M.H. Building "cowshed cultures": A cultural perspective on the promotion of stockmanship and animal welfare on dairy farms. *J. Rural Stud.* **2012**, *28*, 174–187. [CrossRef]
48. Nettle, D.; Bateson, M. The evolutionary origins of mood and its disorders. *Curr. Biol.* **2012**, *22*, R712–R721. [CrossRef]
49. Bethell, E.J. A "How-To" Guide for Designing Judgment Bias Studies to Assess Captive Animal Welfare. *J. Appl. Anim. Welf. Sci.* **2015**, *18*, S18–S42. [CrossRef]
50. Zippia. Available online: www.zippia.com (accessed on 15 August 2023).

animals

MDPI

Article

Assessing the Welfare of Captive Group-Housed Cockroaches, *Gromphadorhina oblongonota*

Danielle Free [1] and Sarah Wolfensohn [2,*]

[1] Marwell Wildlife, Winchester SO21 1JH, UK; danif@marwell.org.uk
[2] School of Veterinary Medicine, University of Surrey, Guildford GU2 7AL, UK
* Correspondence: s.wolfensohn@surrey.ac.uk

Simple Summary: Invertebrate welfare is gaining attention, especially with the rise of insect farming for sustainable food production. Traditional individual welfare monitoring is impractical for large groups, especially when individual identification is difficult. This study adapts the Animal Welfare Assessment Grid (AWAG) for group-level assessments and successfully applies it to a captive group of male *Gromphadorhina oblongonota*. This modified AWAG evaluates welfare based on 12 factors tracked over time, revealing environmental and social factors' impact on *G. oblongonota* welfare. These findings guide practical improvements in care and offer an efficient method to assess invertebrate welfare at the group level.

Abstract: The welfare of invertebrates under human care is of growing concern, particularly with the increasing interest in insect farming as an environmentally sustainable means of producing food. Additionally, individual welfare monitoring systems can be time-consuming and impractical for larger groups, particularly when individual animals are difficult to identify. It is, therefore, imperative to develop a validated system for monitoring terrestrial invertebrate welfare at a group level. The Animal Welfare Assessment Grid (AWAG) is an objective welfare-monitoring tool that has been approved for use with a wide range of species. This study modified the AWAG for large group-level welfare assessments and successfully trialled it on a terrestrial invertebrate species, a group of captive male *Gromphadorhina oblongonota*. The modified template evaluated the group's welfare by scoring changes to 12 factors that could be tracked over time. The results highlight that the welfare of *G. oblongonota* is likely to be influenced by environmental and social factors, and inform practical improvements in *G. oblongonota* care that will result in improved welfare. The findings also demonstrate an efficient way to assess the welfare of invertebrates at the group level, and given the recent UK legislation (Animal Welfare (Sentience) Bill, 2022) plus the emerging interest in invertebrate farming, our findings hold timely significance.

Keywords: Animal Welfare Assessment Grid (AWAG); group welfare; invertebrate; sentience; cockroach; quality of life; welfare

1. Introduction

The welfare of invertebrates under human care is of growing concern [1–4]. Invertebrates are extensively used in laboratories, as 'live food' (to feed non-human animals, e.g., pets, zoo animals and livestock), as pets, for educational purposes (e.g., in classrooms, zoos and museums), and for pest control. There is also increasing interest in farming invertebrates for human consumption (human entomophagy) [5]. There are no definitive figures for the number of invertebrates currently in captivity in the UK, and what data are available typically refer to invertebrates by weight (e.g., kilograms or tonnes), but they likely number in the billions. For example, in 2021, 6000 tonnes of UK-reared insect meal were used in feed for fish farms, pigs and poultry alone. The World Wildlife Foundation (WWF) predict that, by 2050, UK-reared insect meal could be increased to 237,000 tonnes [6].

With such high numbers of invertebrates involved, the importance of assessing invertebrate welfare cannot be underestimated.

Unlike vertebrates, invertebrates receive little protection under UK law. Following the UK's withdrawal from the European Union, Article 13 of the Treaty on the Functioning of the European Union (TFEU), which stated that animals are sentient beings and conferred a duty on member states to pay full regard to animal welfare when formulating and implementing policy [7], was not carried over into UK law. Motivated by public concern and moral consideration for animal welfare, the UK Government passed the Animal Welfare (Sentience) Bill in 2022. Following the provision of scientific evidence of sentience, some invertebrates (e.g., octopuses and lobsters [8]), were included in the bill; however, sentience for the majority of invertebrate species is still not recognised. 'Sentience' is defined by the Farm Animal Welfare Council (FAWC), 2018, as: '*. . .the capability to experience pain, distress and harm*' [9]. Although scientists are still unable to prove sentience for most invertebrate species [10], many agree that, as sentience cannot be disproved, the precautionary principle should be employed; that is, invertebrates should be given the benefit of the doubt with regard to their ability to suffer [1,3,11–13].

Given the lack of legislation, there is a lack of guidance on how to care for invertebrates in captivity. Most available guidance focusses on abiotic conditions. For example, most invertebrates are ectothermic and maintaining a suitable temperature for these species in captivity is key. Lack of guidance also stems from a deficiency of behavioural and ecological data, such as species-specific reproductive and feeding behaviours. This paucity of data is exacerbated by the vast number and diversity of invertebrate species, resulting in 'trial and error' husbandry [4,14], which, if successful for survival and breeding, can result in folklore husbandry, where husbandry methods persist because they have always been used [15].

These hurdles have led to the welfare of invertebrates often being overlooked by society. The World Association of Zoos and Aquariums (WAZA) defines animal welfare as '*. . .a state that is specific for every individual animal; it is how the animal experiences its own world and life through its association with pleasant experiences specific for that species. . .or unpleasant experiences*' [16]. There are, however, many definitions of the term 'welfare', and growing arguments there is no single definition that is all-encompassing [17], with some arguing that the term can only apply to sentient living organisms [18,19]. Although welfare indicators for invertebrates in general are lacking, welfare has been well-studied for vertebrates, from which we can generalise that certain elements of captivity could negatively impact invertebrate welfare. Moreover, as more is understood about sentience in invertebrates and their ability to feel pain, it may be advantageous to employ the precautionary principle and afford invertebrates the benefit of the doubt until confirmed otherwise. Following the precautionary principle, however, and considering the colossal number of individuals at risk, it is important that industries involving living invertebrates identify methods for assessing their welfare [13].

In the UK, legislation discourages the feeding of live vertebrate prey to predators (Animal Welfare Act 2006 and Zoo Licensing Act 1981); however, this legislation is not extended to invertebrate species, with zoos both buying in and breeding invertebrates on site for 'feeding out'. The feeding of live invertebrate prey is believed to be enriching and beneficial for the welfare of the predator, with the welfare of the invertebrates only recently becoming a concern [20–22]. Invertebrates in zoos, particularly those bred for feeding out, are often managed in large numbers in comparatively small enclosures (personal experience). Due to the need to assess multiple individuals and time intervals, assessing the welfare of large groups can be time-consuming, costly and impractical [23,24]. It is especially difficult to individually assess species that cannot be easily identified at an individual level and this is further compounded by the use of larger and more complex environments to house animals in zoos and aquariums [24], highlighting the need for a practical and valid group-level welfare assessment tool.

This exploratory study adapts the Animal Welfare Assessment Grid (AWAG) designed by Narshi et al. [25] to assess its efficacy as a group-level welfare monitoring tool for terrestrial invertebrate species. The AWAG is a practical online cloud-based software [26] devised to assess and monitor the welfare and cumulative lifetime experience of animals. The AWAG tool encompasses the five domains of welfare (see Mellor et al. [27]) across the following four parameters:

Physical: assesses an animal's clinical health, including factors such as body condition, illness and injury.

Behavioural/Psychological: assesses an animal's mental wellbeing and includes factors such as behavioural response to stressors and how often these are encountered. Animals cannot verbally communicate their emotions; therefore, behaviour is used as an indicator to explore their psychological health.

Environmental: assesses the animal's environment, whether it is both suitable and complex, options for social opportunities, and with choice and comfort.

Procedural: assesses how the animal responds to clinical and husbandry events, and includes factors such as handling, changes in routine, and pain from veterinary or management procedures.

The AWAG is unique in that it considers the lifetime experience of the animal, and the cumulative suffering that can impact quality of life. The tool provides a mean score for factors in each parameter and plots these on a grid to create a minimum convex polygon, the area of which is the cumulative welfare assessment score (CWAS) for that moment in time. The CWAS can then be tracked across the animal's lifetime to assess quality of life, allowing for the user to quantify welfare and assess whether treatment or changes in management systems are required or have been successful in improving welfare. The AWAG allows for the user to drill down and identify which factors are positively or negatively affecting welfare and make focused interventions. This tool has been used for a variety of species and various environments, including zoos, farms, companion animal care, and research laboratories [25,28–32]; however, use with large groups of terrestrial invertebrates has yet to be trialled.

2. Materials and Methods

This study modified the AWAG designed by Narshi et al. [25] for assessing the welfare of decapod and cephalopod invertebrates for use with a captive group of *Gromphadorhina oblongonota*, as a proxy for large groups of terrestrial invertebrates housed in captivity. We reviewed the scientific literature for validated welfare indicators for the species on which to base the welfare assessment; however, the literature specific to *G. oblongonota* was scarce, and with over 4000 species of cockroach inhabiting vastly different environments worldwide, extrapolating information from some of the better-understood species, including those regarded as domestic pests or those used in laboratories, was difficult. Where information was available, it largely concerned eradication methods, social behaviour, allergens, neurophysiology, and endocrinology, not welfare indicators.

As per Free et al. [33], we compensated for the lack of peer-reviewed literature by gathering information from commercial and hobbyist communities (via communication with experts through social media and e-mail, and reviewing associated websites, forums, and social media groups), other species within the genus, and direct behavioural observations.

2.1. Study Subjects

The subjects of this study were a zoo-housed group of 31 male *Gromphadorhina oblongonota* at the zoological institution Marwell Zoo (UK). Species of the genus *Gromphadorhina* are managed by many zoos as food for omnivorous and insectivorous species, as well as for educational purposes. The group monitored in this study were housed on-show for viewing by zoo guests. *G. oblongonota* (Figures 1 and 2), also known as the wide-horned hisser, are one of the largest cockroach species in the world, measuring between 5 cm and 10 cm (adult), and are found in southern Madagascar. They are flightless and produce a hissing sound by expelling air through modified spiracles [34–36].

Figure 1. *Gromphadorhina oblongonota* (Marwell Wildlife).

Figure 2. *Gromphadorhina oblongonota* (Kilroy, D).

The group is all-male to prevent overpopulation and is supplemented with males removed from a breeding population used at the zoo for feeding-in. We chose this group to eliminate the need to assess welfare for different life stages and sexes, which have different environmental requirements and a different diet to adult males [35,37,38]. The group were housed in a glass vivarium (Figure 3) measuring 60 × 45 × 90 cm (L × D × H) with a secure mesh upper panel and hinged door for the front. The mesh upper panel allowed for ventilation, lighting (12% Arcadia T5, 6500 K plant light on a seasonal 13:11–11:13 cycle), some UV (low level < 2.0 UVI) and heating (overhead flood halogen 120 W, plus indirect heat from UV/plant lighting). The enclosure had a deep layer of substrate with regular additions of natural leaflitter, maintained by a clean-up-crew of isopods, so additional cleaning was not required. Branching and natural-looking rock facades on two sides of the enclosure completed the environment. The enclosure was misted twice daily to increase the humidity and provide drinking water. Food was typically replaced twice a week and consisted of seasonal browse, fungi and some produce. The *G. oblongonota* were only handled when being added to the enclosure (after removal from the separate breeding colony) or once deceased.

Figure 3. *Gromphadorhina oblongonota* enclosure, Marwell Zoo (Kilroy, D).

2.2. Experimental Design

We assessed the factors utilised by Narshi et al. [25] for their relevance for *Gromphadorhina oblongonota* and ability to be scored accurately at the species and group level, removing those that were not suitable. Following this process, we adapted the remaining scoring definitions by combining elements from previously tested group AWAGs for other species [29] and species-specific criteria for *G. oblongonota*. We detail the final factors below.

Physical (Table 1): The physical parameter considered three animal/outcome-based factors: general condition, presence of injury and activity level. 'General condition' assessed the appearance of the carapace (dull/shiny), body morphology (with both shrivelled and swollen abdomens a concern), difficulty moulting and density of mites [39]. *Gromphadorhina* spp. act as hosts for the mite *Androlaelaps schaeferi* (previously named *Gromphadorholaelaps schaeferi*), and are commensal, feeding on the same food as their host rather than the host itself; therefore, they are not of concern [39–41]. As *A. schaeferi* are primarily found in groups around the spiracles and between the legs of the cockroach [40], they are easy to identify. We merged 'Presence of injury' and 'Observable clinical signs' from Narshi et al. [25], as there is insufficient information in the literature on recognised clinical symptoms for *G. oblongonota*. 'Activity level' simply looked at the percentage of individuals that showed signs of activity during the observation period: 9.30 a.m.–4.30 p.m. We considered activity during the day an indicator of reduced welfare due to the impact on circadian rhythm [42] and the evidence that sleep deprivation in cockroaches and other invertebrates impacts memory formation [43] and can lead to increased metabolic rate, with severe sleep deprivation resulting in death [44]. 'Food intake' was removed, as the amount and type of items fed were too variable to determine accurate changes in the quantity consumed (there was always a large amount remaining when fed again), plus, as food was available 24/7 and *G. oblongonota* are nocturnal [36,45,46], they were rarely seen feeding.

Table 1. Scores for assessing factors within the physical parameter, adapted from Narshi et al. [25].

Score	General Condition	Presence of Injury	Activity Level
	Reduction in general condition would include: dull carapace, difficulty moulting, abnormal body morphology (abdomen not shrivelled/swollen), high density of mites.	Presence of injury including: damaged or missing limbs, tarsi or antennae, lameness/abnormal locomotion, prolapse.	Activity level, e.g., lethargy or hyperactivity. Consider circadian rhythm, which is normally high at night.
1	All of group demonstrated ideal physical condition.	No observable signs of injury.	All of the group demonstrated normal activity levels.
2	1–10% had worse than optimum physical condition.	1–10% had observable signs of injury.	1–10% did not exhibit normal activity levels.
3	11–20% had worse than optimum physical condition.	11–20% had observable signs of injury.	11–20% did not exhibit normal activity levels.
4	21–30% had worse than optimum physical condition.	21–30% had observable signs of injury.	21–30% did not exhibit normal activity levels.
5	31–40% had worse than optimum physical condition.	31–40% had observable signs of injury.	31–40% did not exhibit normal activity levels.
6	41–50% had worse than optimum physical condition.	41–50% had observable signs of injury.	41–50% did not exhibit normal activity levels.
7	51–60% had worse than optimum physical condition.	51–60% had observable signs of injury.	51–60% did not exhibit normal activity levels.
8	61–70% had worse than optimum physical condition.	61–70% had observable signs of injury.	61–70% did not exhibit normal activity levels.
9	71–80% had worse than optimum physical condition.	71–80% had observable signs of injury.	71–80% did not exhibit normal activity levels.
10	>80% had worse than optimum physical condition.	>80% had observable signs of injury.	>80% did not exhibit normal activity levels.

Psychological (Table 2): The psychological parameter measured three factors: abnormal behaviour, response to guest presence and social interaction. 'Abnormal behaviour' considered excessive hissing [38] and increased speed of movement. We changed 'Response to social disruption' to 'Response to guest presence', firstly due to the little interaction the keepers had with the cockroaches, and secondly as the tank was situated in an area of the zoo that can become busy and noisy with guests, who can approach close to the tank. 'Social interaction' was added as an additional factor due to the abundance of the published literature relating to the hierarchal and aggressive behaviour exhibited by male *Gromphadorhina* spp. [47–50]. 'Routine management' from Narshi et al. [25] was removed, as little routine husbandry, other than misting and food input/removal, was carried out, and these were scored under the procedural parameter instead.

Table 2. Scores for assessing factors within the psychological parameter, adapted from Narshi et al. [25].

Score	Abnormal Behaviour	Response to Guest Presence	Social Interaction
	Abnormal behaviour such as moving fast and excessive hissing. Remember to consider normal circadian rhythm when making the assessment.	Reaction to people approaching glass or sudden elevations in noise level. Reactions include: increased startling and hissing.	Evidence of aggression (threat displays or combative) or defensive/submissive behaviours or, e.g., abdominal flick, push, butt or lunge, abdominal extension, abdominal thrash, agonistic hiss and stilt stance [47].

Table 2. *Cont.*

Score	Abnormal Behaviour	Response to Guest Presence	Social Interaction
1	No abnormal behaviours observed in any of the group.	None of the group reacted to guest presence.	No negative social interactions witnessed, either aggressive or defensive.
2	1–10% exhibited abnormal behaviours.	1–10% reacted to guest presence.	1 mild incident witnessed.
3	11–20% exhibited abnormal behaviours.	11–20% reacted to guest presence.	Multiple mild incidences between 2 individuals.
4	21–30% exhibited abnormal behaviours.	21–30% reacted to guest presence.	1 moderate incidence witnessed.
5	31–40% exhibited abnormal behaviours.	31–40% reacted to guest presence.	Multiple mild incidences between >2 individuals.
6	41–50% exhibited abnormal behaviours.	41–50% reacted to guest presence.	Multiple moderate incidences between 2 individuals
7	51–60% exhibited abnormal behaviours.	51–60% reacted to guest presence.	1 severe incidence witnessed
8	61–70% exhibited abnormal behaviours.	61–70% reacted to guest presence.	Multiple severe incidences between 2 individuals
9	71–80% exhibited abnormal behaviours.	71–80% reacted to guest presence.	>80% of individuals involved in at least 1 mild incident.
10	>80% or more of the group exhibited abnormal behaviours.	>80% reacted to guest presence.	>50% of individuals involved in at least 1 moderate-severe incident.

Environmental (Table 3): The environmental parameter measured five factors: environment, group size and structure, enclosure complexity, nutrition, and contingent events. Most of these factors are resource/input-based, analysing what resources the cockroaches have been provided or what they can access, as well as the subsequent effects of these on the cockroaches' overall welfare. Although resource/input-based factors do not account for whether the animals use the provided resources, the lack of validated welfare indicators for invertebrates required us to utilise all means that were available [33]. The factor 'Environment' considered various environmental elements, for example, light, light cycle, UV and cleanliness. Due to their being ectotherms, the temperature, humidity and ventilation in the enclosure was of particular concern [14,35], but ambient noise level and guest proximity were also considered. 'Group size and structure' was essential to assess since overcrowding can lead to disturbances in circadian rhythm, a greater frequency of aggression, reduced growth rates and higher mortality [37,42]. Due to the inability to observe the cockroaches when they were most active, and thus be able to assess the quality and quantity of the natural behaviour that was exhibited, we evaluated 'Enclosure complexity', which focussed on the complexity of the enclosure features and layout, e.g., branching and substrate, and the behaviours that these features enabled the cockroaches to exhibit, e.g., foraging, burrowing, and successful moulting, instead. 'Nutrition' assessed the components of the diet in relation to wild and captive diets and considered that diet presentation likely resulted in feeding and foraging behaviours. Finally, 'Contingent events' considered the impact of irregular events, e.g., the addition of new individuals to the enclosure, guest events or nearby building work, on welfare. For the most part, these scores remained constant as the environment did not change throughout the study. We did not include 'Water quality' and 'Accessibility' from Narshi et al. [25], as neither were relevant.

Table 3. Scores for assessing factors within the environmental parameter, adapted from Narshi et al. [25].

Score	Environment	Group Size and Structure	Enclosure Complexity	Nutrition	Contingent Events
	Suitability of the environment for the species, e.g., location in the zoo, guest viewing, temperature (22–28 °C plus gradient), humidity (60–80%), space, lighting and light cycle, ventilation (no drafts), clean (no mould), drainage, low noise levels.	Considering the number of individuals, group structure, and density in the enclosure.	The enclosure simulates the natural habitat, including branches, bark, leaf litter, rock crevices, among other dark, damp locations, with various heights. Suitable substrate is offered, such as organic soil and leaf litter and natural forage options.	Refers to diet and forage, and presentation.	Contingent events include: enclosure changes, building works, guest events/educational aids, bin collection, deliveries.
1	The enclosure is ideal for the species.	Group size in accordance with natural group size; group structure is appropriate; suitable density for the enclosure size.	Ability to demonstrate all natural behaviours.	Nutrition is optimally suited to the species and individual. (nutritional, physiological and behavioural).	None.
2	One factor is below average.	Group structure is marginally different from appropriate group structure.	The ability to demonstrate natural behaviours, but the available options for this are not ideal.	Nutrition available has a marginally decreased appropriateness to accommodate species-specific needs.	Event outside of the enclosure (e.g., continuing construction work) taking place with little disruption
3	Two/three factors are below average.	An increased or decreased number of animals present in comparison to the natural group size range, no overstocking.	The ability to demonstrate natural behaviours, but there are few possibilities for this.	Nutrition available has a moderately decreased appropriateness for species-specific needs.	Event outside of the enclosure (e.g., continuing construction work) with slight disruption, e.g., noise or vibrations
4	Four/five factors are below average.	Group structure is moderately different from appropriate group structure.	The ability to demonstrate one form of natural behaviour is constrained due to the enclosure design.	Nutrition available is largely inappropriate to accommodate species-specific needs.	An external event that causes some disruption OR a change in the enclosure's contents without any other events occurring.
5	Six factors below average.	A somewhat greater animal density than suitable for enclosure size (many young present without a decrease in the adult population)	A certain type of natural behaviour is unable to be demonstrated because the option is not offered.	Nutrition available is inadequate to fulfil behavioural needs of the species.	External incident that causes a visible interruption OR movement into a familiar environment without any other events occurring.
6	Seven factors below average.	An increased or decreased number of animals present in comparison to the natural group size range, with marginal overstocking.	The options to demonstrate natural behaviours are limited, preventing the demonstration of particular natural behaviours associated with enclosure design.	Nutrition available is inadequate to fulfil physiological needs of the species.	External incident that causes a visible interruption AND movement into a familiar environment.

Table 3. *Cont.*

Score	Environment	Group Size and Structure	Enclosure Complexity	Nutrition	Contingent Events
7	Eight factors below average.	Group structure is greatly different from appropriate group structure.	The options to demonstrate natural behaviours are limited, preventing the demonstration of multiple natural behaviours associated with enclosure design.	Nutrition available is inadequate to fulfil the behavioural and physiological needs of the species.	Movement into unfamiliar enclosure OR addition of unfamiliar animal.
8	Nine factors below average.	Animal density is moderately greater than suitable for enclosure size.	The options to demonstrate natural behaviours are limited, preventing the demonstration of the majority of natural behaviours associated with enclosure design.	Available nutrition is inadequate to fulfil behavioural, physiological and nutritional needs of the species.	Movement into unfamiliar enclosure AND addition of unfamiliar animal.
9	Ten factors below average.	Animal density is much greater than enclosure can accommodate.	The options to demonstrate natural behaviours are very limited, preventing the demonstration of virtually all of the natural behaviours associated with enclosure design.	No nutrition is available.	External incident that causes a definite interruption AND movement into unfamiliar enclosure
10	All factors scored are inadequate—the enclosure is inappropriate for the species being monitored.	Group size greatly differed from natural group size or a significant level of overstocking.	The animal is unable to demonstrate natural behaviours associated with enclosure design because the options are not offered.	Nutrition is dangerous for the species.	Mixture of events: extended external incident, movement into an unfamiliar enclosure, introduction of unfamiliar animals. Extreme detrimental levels of disruption.

Procedural (Table 4): The procedural parameter evaluated one factor: the effect of intervention. We removed 'Isolation/Restraint', 'Impact of veterinary procedures', 'Change in daily routine' and 'Sedation/Anaesthesia' from Narshi et al. [25], as veterinary interventions were very unlikely to occur for this group and daily routine was variable anyway.

Table 4. Scores for assessing factors within the procedural parameter, adapted from Narshi et al. [25].

Score	Effect of Intervention
	For example, removing old food or dead individuals, changes to environmental complexity, etc. Stress-related behaviours include: hissing, burrowing or fast movement away from the disturbance
1	No intervention.
2	1–10% reacted to the intervention by exhibiting stress-related behaviours.
3	11–20% reacted to the intervention by exhibiting stress-related behaviours.
4	21–30% reacted to the intervention by exhibiting stress-related behaviours.
5	31–40 reacted to the intervention by exhibiting stress-related behaviours.
6	41–50% reacted to the intervention by exhibiting stress-related behaviours.
7	51–60% reacted to the intervention by exhibiting stress-related behaviours.
8	61–70% reacted to the intervention by exhibiting stress-related behaviours.
9	71–80% reacted to the intervention by exhibiting stress-related behaviours.
10	>80% reacted to the intervention by exhibiting stress-related behaviours.

We modified many of the physical, psychological and procedural parameter factors to adapt the AWAG template designed by Narshi et al. [25] for group-level assessment. For six of the factors, this involved using a percentage of the group for each definition, from 1 to 10. In most cases, a score of 1 indicated that the whole group had ideal welfare for that factor, decreasing by 1–10% for scores 2–9, and a score of 10 indicated that more than 80% of the group had the poorest welfare for that factor. Using a percentage range when scoring allowed for those individuals that could not be seen at each observation.

'Social behaviour' included a mixture of number of aggressive interactions, severity of interaction and the percentage of the group involved in interactions. The five environmental factors did not require adaptation to remain relevant for a group.

The final AWAG consisted of 12 factors within the four parameters, with each scored incrementally on a scale from 1 (best welfare state) to 10 (worst welfare state). We carefully defined each score to enhance interscorer agreement and, during preliminary observations, baseline scores were collectively agreed on by two independent observers, who also scored the group's welfare simultaneously on two occasions during these observations. Interscorer agreement was then determined by calculating the percentage of scores that varied between the two scorers and was calculated at 100%. The observers were undergraduate students of relevant disciplines that were trained by the primary author. See Justice et al. [29] and Brouwers and Duchateau [51] for further detail on the methods used. The score definitions are presented in Tables 1–4.

2.3. Welfare Analysis

Discussions with the keeping team and four preliminary observations across two days allowed a baseline score for the group to be established. We counted injuries during this period, as each individual was seen at least once. Five injuries were identified and fed into the baseline score, which only changed if additional injuries were noted, or an injured individual died. As it was not always possible to see all the individuals at each observation without disruption which would negatively impact welfare, those out of sight were scored at the baseline level.

Two or three welfare assessments were conducted per day between 09:30 and 16:30, between 11 and 21 September 2023. Multiple assessments were conducted per day to detect differences in the group's welfare over the course of a day as they experienced changes in temperature, humidity, presence/absence of food, guest numbers etc. Multiple assessments also increased the ability to see all individuals. The group was monitored for 15 min per observation and scored using the modified AWAG. The scores were input into the AWAG software (https://awag.org.uk/, accessed on 23 September 2023) [26] for analysis. The AWAG cloud-based software generates an average for each parameter and plots this as a minimum convex polygon graph for every welfare assessment. The Cumulative Welfare Assessment Score (CWAS) is derived from the area of the polygon [25,29,31]. Temperature in the tank was recorded using a DS1921G-F5 Thermochron® iButton® [52], which noted the temperature in degrees Celsius every 2 h.

The study received the University of Surrey's and Marwell Wildlife's ethical approval prior to data collection.

3. Results

The total number of individuals in the enclosure was 31 at the start of the data collection period and 30 at the end due to the death of one individual from natural causes.

Twenty-three observations were conducted across eight days. The average number of individuals seen per observation was 27 (max. 31, min. 21) due to leaflitter providing some cover but not enough for all the individuals present (the maximum out of sight during an assessment was 10; the average number out of sight was 4). Changes were detected in welfare score over this period and are visually presented in Figures 4–8. Figure 4 depicts the change in the group's CWAS over time. The maximum score that can be attained for a single assessment is 200. The maximum CWAS score for the group during the study was 13.3, whilst the minimum was 6.1 (Figure 4).

Figure 4. The CWAS for the group over the study period. The lower the score, the better the welfare. This form of visual representation highlights specific events that may be impacting welfare, in addition to changes in the trend of welfare over time, with a rising line indicating welfare is declining or a falling line indicating welfare is improving. Annotations identify which factor scores changed as a result of specific events, indicating a decline (peak) or improvement (trough) in group welfare.

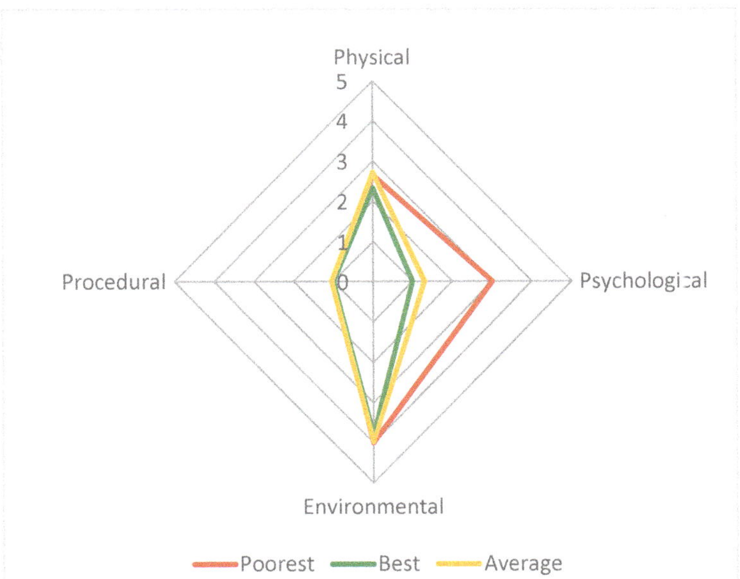

Figure 5. This radar chart compares the four parameter scores (physical, psychological, environmental, and procedural) for three welfare assessments: 13 September 15:30 (poorest group welfare score—13.3, red), 21 September 10:30 (best group welfare score—6.1, green) and the average for the study period (7.9, orange), on a scale from 1 to 10, with 1 being the best possible score and 10 the poorest. The axes in this figure are adjusted to improve readability.

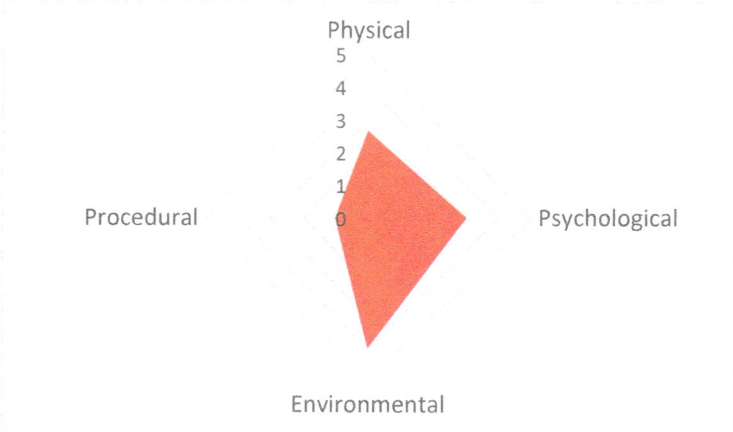

Figure 6. The welfare polygon for 13 September 15:30. The environmental parameter has the highest score (4), resulting from high scores for the factors: environment (5), group size and structure (6) and enclosure complexity (6). The psychological parameter is also high due to the factor of social interaction (5). The axes in this figure are adjusted to improve readability.

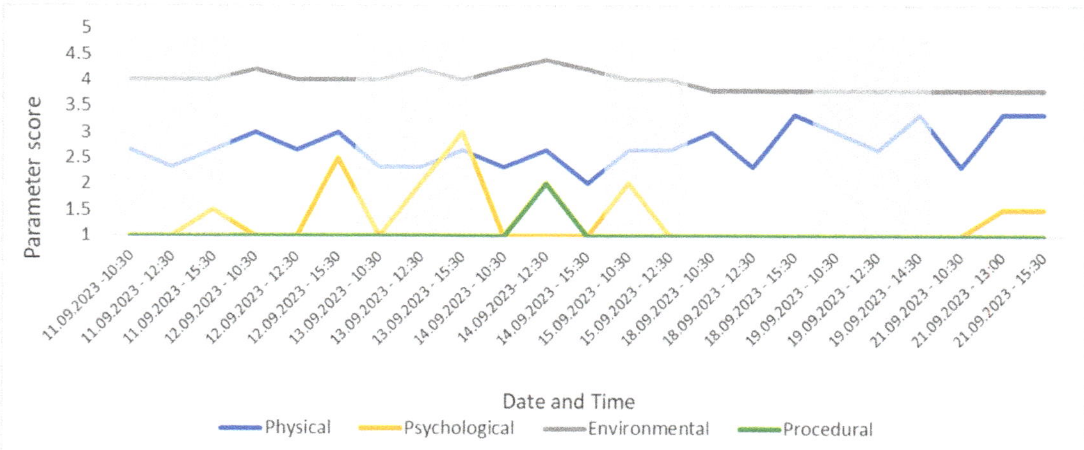

Figure 7. The four parameter scores for the group, presented over time for the study period. This figure shows changes in welfare related to specific parameters, allowing for the assessor to easily drill down into the CWAS.

The poorest welfare score for the group (13.3) occurred on 13 September at the 15:30 assessment, whilst the best welfare score for the group (6.1) occurred on 21 September at the 10:30 assessment. The average welfare score for the group for the study period was 7.9 (Figure 5). The psychological parameter had the greatest impact on the CWAS for the 13 September 15:30 (Figure 6 and Table 5). The factors of environment, group size and structure, and enclosure complexity did not decline below scores of 5, 6 and 5, respectively, across the study period.

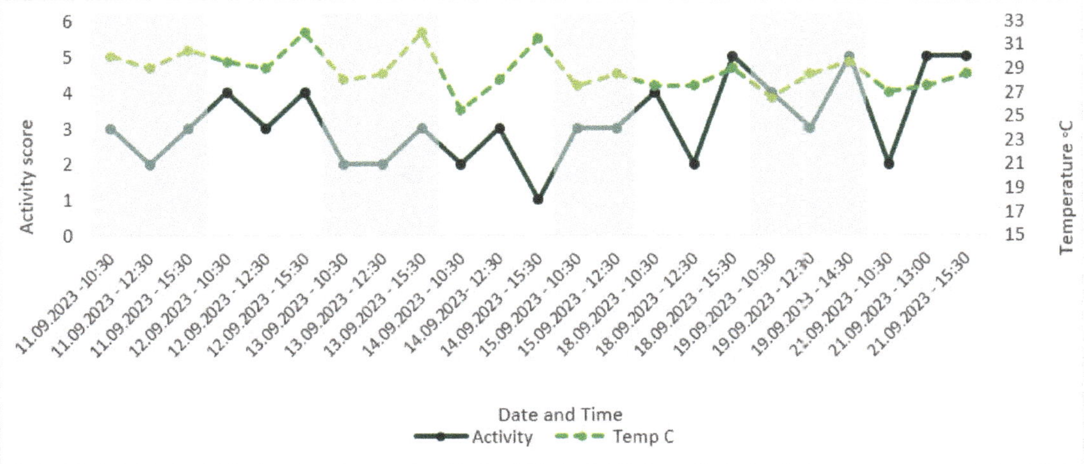

Figure 8. Comparison of score for the factor 'activity' over time, with corresponding enclosure temperature (°C). Activity during the day was deemed an indicator of poor welfare; therefore, an increase in score for activity represented an increase in group activity level.

Table 5. Factor scores for 13 September 15:30. Highlighted in red are the three factors that resulted in the environmental parameter having the highest score for that assessment and the single factor that raised the psychological parameter.

Parameter/Factor	Score	Parameter/Factor	Score
Physical		**Environmental**	
General condition	2	Environment	5
Presence of injury	3	Group size and structure	6
Activity level	3	Enclosure complexity	6
Psychological		Nutrition	2
Response to guests	1	Contingent events	1
Social interaction	5	**Procedural**	
		Effect of intervention	1

Changes to the four parameter scores can also be viewed across time. The psychological parameter exhibits the greatest change in score (min. 1–max. 3). The environmental parameter remains fairly consistent throughout the study and the procedural parameter deviates from 1 only once (14 September 23 12:30), due to the removal of old food by a keeper during the observation (Figure 7).

Temperatures remained above the minimum of 21–24 °C, as recommended in the husbandry-related literature throughout the study period, with the lowest recorded temperature of 25.5 °C [41,53]. The maximum temperature recorded was 32 °C, exceeding the maximum recommended temperature of 27–30 °C [41,53]. CWAS visually followed a similar trend to time of day (Figure 4). A Spearman's rank correlation was computed using R [54] to assess the relationship between CWAS and time of day and found a weak but insignificant positive correlation, $rs(21) = 0.36$, $p = 0.094$. Activity level appeared to follow the same trend as temperature (Figure 8), and a Spearman's rank correlation [54] found a very weak insignificant positive correlation between the two variables, $rs(21) = 0.11$, $p = 0.607$.

4. Discussion

The aim of this study was to expand the work by Narshi et al. [25] to examine whether the previously validated Animal Welfare Assessment Grid (AWAG) could be modified to objectively assess the welfare of a large group of terrestrial invertebrates. Through adapting the welfare scoring criteria for *Gromphadorhina oblongonota* at the group level and successfully trialling it with a group of 31 males at Marwell Zoo, we provide evidence the AWAG can be utilised to monitor the welfare of terrestrial invertebrates at the group level. With the recent passing of the Animal Welfare (Sentience) Bill into UK legislation in 2022, and the resulting questions and concerns for invertebrate welfare that this raised, in addition to the growing interest in farming invertebrates for human consumption, this is a timely result.

Although welfare is a subjective experience, assessing welfare at the group level leaves individual differences and individual personality unaccounted for. Group-level welfare assessments tend to focus more on what the animal has been provisioned with, harking back to the 'Five Freedoms', rather than assessing animal-based measures such as individual behaviour and physical and psychological health. However, through using the AWAG model, we included animal-based measures in this group-level welfare assessment and our results indicate that, through using this tool, changes in welfare and welfare trends over time can be identified and tracked at the group level. The results also show that it is possible to identify which factor(s) may be impacting welfare, allowing for animal care givers to focus on improving these. For *G. oblongonota,* our results suggest the environment, group size and structure, and enclosure complexity may impact baseline welfare score. In addition, the factors 'Activity level' under the physical parameter and 'Social interaction' under the psychological parameter varied the most between observations, and thus had the greatest impact on changes in welfare score over the course of this study.

4.1. *The Impact of Environmental Factors on Welfare Score*

As they are ectotherms, low temperatures can have a negative impact on the welfare of cockroaches. Low temperatures, i.e., 8–10 °C, can induce a 'chill-coma', where cockroaches lose mobility but can survive if temperatures are reversed before a chill-injury occurs [55]; therefore, it is important to monitor enclosure temperature closely. Although maximum temperatures of 27–30 °C are recommended in the literature [41,53], the temperature in *G. oblongonota's* wild habitat in southern Madagascar can reach 40 °C. Except for increases in reproductive activity [41,53,55] and increased metabolic rate [56], we found no evidence in the literature for a negative impact of higher temperatures on welfare. Even so, we increased the welfare score when temperatures were above the recommended level in case, over time, correlations were found with other factors.

The majority of cockroach species are negatively phototactic, meaning that they will move away from light. In this enclosure, there were very few options to move away from or seek shelter from the light, which was on a seasonal 13:11–11:13 cycle. Leaflitter provided some cover but not enough for all the individuals present. The remainder appeared to choose locations in the enclosure with as much cover as possible, either in the shallow cervices on the rock facades (Figure 9) or on the underside of the branches (Figure 10).

We noted anecdotally that when leaflitter was introduced, activity levels increased and many of the cockroaches moved to burrow into the new substrate. For negatively phototactic species, being unable to escape light can lead to increased stress and anxiety [57–59], negatively impacting welfare and resulting in lower survival rates [42].

Cockroaches are also positively thigmotaxic; that is, they will actively seek contact with objects in their environment. This behaviour increases safety and security, particularly from predators, and is often used as an escape response. Cockroaches tend to avoid open spaces and will preferentially locate themselves close to or touching walls or other vertical surfaces (Figure 9). It is likely that this behaviour is also associated with light avoidance [42]. Studies have shown that thigmotactic deprivation, e.g., lack of access to shelters, can lead to stress, resulting in reduced growth, increased energy consumption, reduced fecundity and reduced survival rate [42,60,61].

Figure 9. Three *Gromphadorhina oblongonota* exhibiting thigmotaxic and negative phototactic behaviour in one of the wall crevices of the enclosure. (Kilroy, D).

Figure 10. *Gromphadorhina oblongonota* in the enclosure at Marwell Zoo. Note the position of all bar one individual on the underside of the branches. (Kilroy, D).

4.2. Other Factors That Impacted Welfare Score

We considered activity during the day to be a sign of compromised welfare due to the disruption of the circadian rhythm. Without comparison with other colonies or evidence for wild individuals, it is unknown whether the amount of activity seen during the day was typical for the species. By highlighting this using the AWAG, however, we were able to track changes over time. As ectotherms, temperature plays a key role in the activity level of *G. oblongonota*, with temperatures below 21 °C leading to reduced activity level and lethargy and high temperatures increasing activity level [41,53,55]. Activity level may therefore be an indicator for the impact of environmental parameters on welfare in this species.

Alongside 'Group size and structure', we identified 'Social interaction' as a key welfare indicator because it has been demonstrated previously that male *Gromphadorhina* exhibit extensive dominance hierarchies [47,49,50,62]. Although cockroaches are typically highly social and aggregate in large numbers, wild *Gromphadorhina* spp. usually live in mixed-sex groups of only ca. 10 individuals, in which males exhibit territoriality, combat behaviours, and dominance hierarchies [35,63]. These social dynamics suggest that high numbers of males in a small area will lead to an increase in these negative social interactions. As this group of 31 males was maintained at a fairly high stocking density, we used the frequency, severity, and number of individuals involved in negative interactions as a welfare indicator [38]. Although negative social interactions remained low in number and severity, and there were only two assessments in which we scored welfare a 4 (12 September 15:30) or 5 (13 September 15:30; Figure 6), it is possible that a stable social hierarchy had already been established prior to the start of the study [47] and that this factor will change more when there are changes to the group (e.g., new individuals are added to the enclosure). We observed some individuals to be missing parts of their antennae and tarsi, which could be a result of previous bouts of aggression or, alternatively, old age [39].

We did not observe any cockroach responses to guests. A lack of responses could have been a result of the assessors' presence changing guest behaviour and leading to fewer negative behaviours such as banging on the glass window or shouting. Fewer responses could also have been because the enclosure was situated behind a glass window that lessened sound and vibrations. Alternatively, stationary behaviour may have continued because there were few hiding places for cockroaches, meaning escaping out of the sight of guests was not possible. Another explanation may be that presence if guests did not negatively impact welfare; however, this seems unlikely considering that the negative impact of guests on animal welfare has been described in detail (reviewed in Sherwen and Hemsworth [64]). Further research is required to determine whether there are more relevant methods for scoring the impact of guests on the welfare of invertebrates.

In addition to the results from this study supporting use of the AWAG for welfare assessments of groups, they will also feed directly into husbandry and enclosure changes for this group of *G. oblongonota*. These changes will include a larger enclosure, focussing on floor space rather than height as *G. oblongonota* are primarily terrestrial [36,65,66]. A greater floor area will also allow for gradients in temperature, humidity and light, providing choice and control over how the environment is used [1,39,53]. Changes will also include the introduction of canopy cover to reduce the amount of light falling on the floor, as would occur in the wild forest habitat, and the addition of more leaflitter and other shelters that can provide escape from the light and areas of higher humidity.

4.3. Limitations

Whilst we are satisfied that this trial was successful at modifying a welfare assessment tool for terrestrial invertebrates at the group level, the number of welfare assessments employed in this study was low. A long-term ($\geq$95 days, as per Justice et al. [29]) evaluation of this modified AWAG scoring template to allow for an improved analysis of welfare trends over time is the next step. Adding nocturnal observations to the methods will allow for us to gain a better understanding of behaviour and whether the results obtained from diurnal observations only are a true representation of welfare. Unless organisations maintain their cockroach colonies in reverse light–dark cycle, a lack of nocturnal observations is likely to be a widespread issue. Once periods of peak activity are identified, observations can occur at those times to reduce the impact of time of day on the results. Another issue we faced was an inability to see all the individuals in the group during each assessment without causing a disturbance. Ideally, the assessment score would represent the total sum of the welfare of all the individuals in the group; however, in practice this can be difficult, so we are presenting a tool that we believe can still positively contribute to our understanding of group-level welfare within the confines of these limitations.

Abnormal behaviour is a potentially important welfare indicator [67–69]. We ended up not scoring it in this study because it was not possible to view the cockroaches at night, when they are most active and thus most likely to exhibit abnormal behaviour. It was also not possible to assess the frequency of hissing due to the enclosure being situated behind an additional glass window, meaning that hisses were not audible to the observers. This is something to address in future studies.

In this study, we modified the AWAG template for *G. oblongonota*. More research is necessary to determine whether a single generalised template could be designed for the assessment of multiple species of terrestrial invertebrates at the group level, taking into account differences in husbandry methods and environment. In addition, we based most factors in this modified AWAG on the percentage of individuals that deviated from the norm. Further research should explore combining percentages with severity (as per the factor 'Social interaction') to improve how representative the scores are for the welfare of the individual within the group. Furthermore, for the parameter averages to be truly representative, scores between 1 and 10 should uniformly increase or decrease. This should be accounted for in future studies.

One of the key steps in designing a welfare assessment tool is reviewing the current literature on animal welfare research and the species or taxa of concern. There are few studies of *G. oblongonota* welfare in captivity, resulting in a heavy reliance on information gathered from the wild concerning environment and behaviour. These data may not always be directly related to their welfare in captivity [70,71]. As a result, it is important to be conscious of the relevance of this information for the captive environment. In some situations, it may be preferable for welfare assessments to not compare captivity to the natural environment and instead concentrate on how well the environment fulfils the behavioural requirements of the species, giving greater weight to animal-based factors [33].

5. Conclusions

This study advances the welfare assessment of large groups of terrestrial invertebrates by adapting the Animal Welfare Assessment Grid (AWAG) for *Gromphadorhina oblongonota*. Our findings underscore its suitability as a practical, easy-to-use tool for objective group-level welfare evaluation. These are essential requirements for a monitoring tool, given that current individual welfare monitoring systems can be time-consuming and impractical for large groups [23,24]. These findings contribute to tangible improvements in the care of *G. oblongonota* in captivity, ultimately elevating their welfare standards but also contributing to the ability to monitor and improve welfare for groups of other invertebrate species.

Based on our findings, we would suggest the following recommendations for others looking to adapt the AWAG for use with terrestrial invertebrates:

- Adapt the AWAG scoring criteria to the species, based on thorough research.
- Improve the representation of individual welfare by combining the percentage of impacted individuals and severity of the impact in the scoring criteria.
- Standardise the time of day the assessment takes place, which should include the species' most active period.
- Assess welfare regularly.
- Utilise the assessment results to drive changes that will improve animal welfare.

Author Contributions: Conceptualization, D.F. and S.W.; methodology, D.F.; validation, D.F.; formal analysis, D.F.; data curation, D.F.; writing—original draft preparation, D.F. and S.W.; writing—review and editing, S.W.; supervision, S.W.; project administration, D.F.; funding acquisition, S.W. All authors have read and agreed to the published version of the manuscript.

Funding: The publication of this work was supported by the University of Surrey.

Institutional Review Board Statement: Ethical review and approval were waived in this study as there was no manipulation to the study animal or their environment.

Informed Consent Statement: Not applicable.

Data Availability Statement: The data presented in this study are available on request from the corresponding author.

Acknowledgments: We thank Nicola Argent for her preliminary attempts at adapting the AWAG for *Gromphadorhina oblongonota* as part of her undergraduate project and students Ella Holding and Catherine Basford for data collection. We also thank the Herptiles Team at Marwell Wildlife for their time and support throughout this study. Finally, we thank those at Marwell Wildlife who reviewed final drafts and the anonymous reviewers for their valuable contributions.

Conflicts of Interest: The authors declare no conflict of interest.

References

1. Cooper, J.E. Invertebrate Care. *Vet. Clin. Exot. Anim. Pract.* **2004**, *7*, 473–486. [CrossRef] [PubMed]
2. Boppré, M.; Vane-Wright, R.I. Welfare Dilemmas Created by Keeping Insects in Captivity. In *The Welfare of Invertebrate Animals*; Animal Welfare, Carere, C., Mather, J., Eds.; Springer International Publishing: Cham, Switzerland, 2019; Volume 18, pp. 23–67, ISBN 978-3-030-13946-9.
3. van Huis, A. Welfare of Farmed Insects. *J. Insects Food Feed.* **2019**, *5*, 159–162. [CrossRef]
4. Delvendahl, N.; Rumpold, B.A.; Langen, N. Edible Insects as Food–Insect Welfare and Ethical Aspects from a Consumer Perspective. *Insects* **2022**, *13*, 121. [CrossRef]
5. Pippinato, L.; Gasco, L.; Di Vita, G.; Mancuso, T. Current Scenario in the European Edible-Insect Industry: A Preliminary Study. *J. Insects Food Feed.* **2020**, *6*, 371–381. [CrossRef]
6. WWF-UK. The Future of Feed: A WWF Roadmap to Accelerating Insect Protein in UK Feeds. 2022. Available online: https://www.wwf.org.uk/sites/default/files/2022-06/future_of_feed_full_report.pdf (accessed on 12 September 2023).
7. The Member States. *Treaty of Lisbon Amending the Treaty on European Union and the Treaty Establishing the European Community, Signed at Lisbon, 13 December 2007*; Proactis: Aberdeen, UK, 2007; Volume 306.
8. Birch, J.; Burn, C.; Schnell, A.; Browning, H.; Crump, A. *Review of the Evidence of Sentience in Cephalopod Molluscs and Decapod Crustaceans*; London School of Economics and Political Science: London, UK, 2021.
9. Jinman, P. Animal Sentience 2018. Available online: https://assets.publishing.service.gov.uk/media/5f0f05db3a6f400394d55225/FAWC_letter_on_animal_sentience_16_March_2018.pdf (accessed on 12 September 2023).
10. Lambert, H.; Elwin, A.; D'Cruze, N. Wouldn't Hurt a Fly? A Review of Insect Cognition and Sentience in Relation to Their Use as Food and Feed. *Appl. Anim. Behav. Sci.* **2021**, *243*, 105432. [CrossRef]
11. Sherwin, C.M. Can Invertebrates Suffer? Or, How Robust Is Argument-By-Analogy? *Anim. Welf.* **2001**, *10*, S103–S118. [CrossRef]
12. Knutsson, S.; Munthe, C. A Virtue of Precaution Regarding the Moral Status of Animals with Uncertain Sentience. *J. Agric. Environ. Ethics* **2017**, *30*, 213–224. [CrossRef]
13. Barrett, M.; Chia, S.Y.; Fischer, B.; Tomberlin, J. Welfare Considerations for Farming Black Soldier Flies, Hermetia Illucens (Diptera: Stratiomyidae): A Model for the Insects as Food and Feed Industry. *J. Insects Food Feed.* **2022**, *9*, 119–148. [CrossRef]
14. Saul-Gershenz, L.S.; Arnold, R.A.; Scriber, J.M. Design of Captive Environments for Endangered Invertebrates. In *Conservation of Endangered Species in Captivity: An Interdisciplinary Approach*; Gibbons, E.F., Durrant, B.S., Demarest, J., Eds.; State University of New York Press: New York, NY, USA, 1995; pp. 51–71, ISBN 978-0-7914-1911-3.
15. Arbuckle, K. Folklore Husbandry and a Philosophical Model for the Design of Captive Management Regimes. *Herpetol. Rev.* **2013**, *44*, 448–452.
16. WAZA. Our Approach to Animal Welfare. 2023. Available online: https://www.waza.org/ (accessed on 12 September 2023).
17. Arndt, S.S.; Goerlich, V.C.; van der Staay, F.J. A Dynamic Concept of Animal Welfare: The Role of Appetitive and Adverse Internal and External Factors and the Animal's Ability to Adapt to Them. *Front. Anim. Sci.* **2022**, *3*, 908513. [CrossRef]
18. Duncan, I.J.H. The Changing Concept of Animal Sentience. *Appl. Anim. Behav. Sci.* **2006**, *100*, 11–19. [CrossRef]
19. Duncan, I.J.H. Is Sentience Only a Nonessential Component of Animal Welfare? *Anim. Sentience* **2016**, *1*, 6. [CrossRef]
20. Cooper, J.E.; Williams, D.L. The Feeding of Live Food to Exotic Pets: Issues of Welfare and Ethics. *J. Exot. Pet Med.* **2014**, *23*, 244–249. [CrossRef]
21. Keller, M. Feeding Live Invertebrate Prey in Zoos and Aquaria: Are There Welfare Concerns? *Zoo Biol.* **2017**, *36*, 316–322. [CrossRef]
22. Kagan, R.; Allard, S.; Carter, S. What Is the Future for Zoos and Aquariums? *J. Appl. Anim. Welf. Sci.* **2018**, *21*, 59–70. [CrossRef] [PubMed]
23. Collins, S.; Burn, C.C.; Wathes, C.M.; Cardwell, J.M.; Chang, Y.-M.; Bell, N.J. Time-Consuming, but Necessary: A Wide Range of Measures Should Be Included in Welfare Assessments for Dairy Herds. *Front. Anim. Sci.* **2021**, *2*, 703380. [CrossRef]
24. DiVincenti, L.; McDowell, A.; Herrelko, E.S. Integrating Individual Animal and Population Welfare in Zoos and Aquariums. *Animals* **2023**, *13*, 1577. [CrossRef]
25. Narshi, T.M.; Free, D.; Justice, W.S.M.; Smith, S.J.; Wolfensohn, S. Welfare Assessment of Invertebrates: Adapting the Animal Welfare Assessment Grid (AWAG) for Zoo Decapods and Cephalopods. *Animals* **2022**, *12*, 1675. [CrossRef]
26. Reuben Digital Animal Welfare Assessment Grid. Available online: https://awag.org.uk/ (accessed on 29 September 2023).

27. Mellor, D.J.; Beausoleil, N.J.; Littlewood, K.E.; McLean, A.N.; McGreevy, P.D.; Jones, B.; Wilkins, C. The 2020 Five Domains Model: Including Human–Animal Interactions in Assessments of Animal Welfare. *Animals* **2020**, *10*, 1870. [CrossRef]
28. Honess, P.; Wolfensohn, S. The Extended Welfare Assessment Grid: A Matrix for the Assessment of Welfare and Cumulative Suffering in Experimental Animals. *Altern. Lab. Anim.* **2010**, *38*, 205–212. [CrossRef]
29. Justice, W.S.M.; O'Brien, M.F.; Szyszka, O.; Shotton, J.; Gilmour, J.E.M.; Riordan, P.; Wolfensohn, S. Adaptation of the Animal Welfare Assessment Grid (AWAG) for Monitoring Animal Welfare in Zoological Collections. *Vet. Rec.* **2017**, *181*, 143. [CrossRef] [PubMed]
30. Wolfensohn, S.; Shotton, J.; Bowley, H.; Davies, S.; Thompson, S.; Justice, W. Assessment of Welfare in Zoo Animals: Towards Optimum Quality of Life. *Animals* **2018**, *8*, 110. [CrossRef] [PubMed]
31. Ryan, M.; Waters, R.; Wolfensohn, S. Assessment of the Welfare of Experimental Cattle and Pigs Using the Animal Welfare Assessment Grid. *Animals* **2021**, *11*, 999. [CrossRef]
32. Malkani, R.; Paramasivam, S.; Wolfensohn, S. Preliminary Validation of a Novel Tool to Assess Dog Welfare: The Animal Welfare Assessment Grid. *Front. Vet. Sci.* **2022**, *9*, 940017. [CrossRef] [PubMed]
33. Free, D.; Justice, W.S.M.; Smith, S.J.; Howard, V.; Wolfensohn, S. An Approach to Assessing Zoo Animal Welfare in a Rarely Studied Species, the Common Cusimanse *Crossarchus obscurus*. *J. Zool. Bot. Gard.* **2022**, *3*, 420–441. [CrossRef]
34. Nelson, M.C.; Fraser, J. Sound Production in the Cockroach, *Gromphadorhina portentosa*: Evidence for Communication by Hissing. *Behav. Ecol. Sociobiol.* **1980**, *6*, 305–314. [CrossRef]
35. Schal, C.; Gautier, J.-Y.; Bell, W.J. Behavioural Ecology of Cockroaches. *Biol. Rev.* **1984**, *59*, 209–254. [CrossRef]
36. Monahan, C.F.; Bogan, J.E.; LaDouceur, E.E.B. Histological Findings in Captive Madagascar Hissing Cockroaches (*Gromphadorhina portentosa*) and a Literature Review. *Vet. Pathol.* **2023**, 03009858231166659. [CrossRef]
37. Bell, W.J.; Roth, L.M.; Nalepa, C.A. *Cockroaches: Ecology, Behavior, and Natural History*; JHU Press: Baltimore, MA, USA, 2007; ISBN 978-0-8018-8616-4.
38. Brereton, J. The Behavioural Biology of Invertebrates. In *The Behavioural Biology of Zoo Animals*; Rose, P., Ed.; CRC Press: Boca Raton, FL, USA, 2022; pp. 269–282, ISBN 978-1-00-077657-7.
39. Kandilian, K. *Roach Crossing's Cockroach Husbandry Guide*; Roach Crossing LLC: Royal Oak, MI, USA, 2022.
40. Yoder, J.A.; Barcelona, J.C. Food and Water Resources Used by the Madagascan Hissing-Cockroach Mite, Gromphadorholaelaps Schaeferi. *Exp. Appl. Acarol.* **1995**, *19*, 259–273. [CrossRef]
41. Chua, J.; Fisher, N.A.; Falcinelli, S.D.; DeShazer, D.; Friedlander, A.M. The Madagascar Hissing Cockroach as an Alternative Non-Mammalian Animal Model to Investigate Virulence, Pathogenesis, and Drug Efficacy. *J. Vis. Exp.* **2017**, *24*, 56491. [CrossRef]
42. Retardo-Agua, J.; Crausos, K.; Sasam, J.M. Growth Rate and Thigmotactic Behavior of Turkestan Cockroach (*Blatta lateralis*) under Different Illumination Conditions. *IJRIAS* **2023**, *8*, 129–140. [CrossRef]
43. Michel, M.; Lyons, L.C. Unraveling the Complexities of Circadian and Sleep Interactions with Memory Formation through Invertebrate Research. *Front. Syst. Neurosci.* **2014**, *8*, 133. [CrossRef]
44. Stephenson, R.; Chu, K.M.; Lee, J. Prolonged Deprivation of Sleep-like Rest Raises Metabolic Rate in the Pacific Beetle Cockroach, *Diploptera punctata* (Eschscholtz). *J. Exp. Biol.* **2007**, *210*, 2540–2547. [CrossRef] [PubMed]
45. Mishra, M.; Meyer-Rochow, V.B. Fine structural description of the compound eye of the Madagascar 'hissing cockroach'*Gromphadorhina portentosa* (Dictyoptera: Blaberidae). *Insect Sci.* **2008**, *15*, 179–192. [CrossRef]
46. Warrant, E.J. The remarkable visual capacities of nocturnal insects: Vision at the limits with small eyes and tiny brains. *Philos. Trans. R. Soc. B Biol. Sci.* **2017**, *372*, 20160063. [CrossRef] [PubMed]
47. Clark, D.C.; Moore, A.J. Social Interactions and Aggression among Male Madagascar Hissing Cockroaches (*Gromphadorhina portentosa*) in Groups (Dictyoptera: Blaberidae). *J. Insect Behav.* **1994**, *7*, 199–215. [CrossRef]
48. Clark, D.C.; Moore, A.J. Variation and Repeatability of Male Agonistic Hiss Characteristics and Their Relationship to Social Rank in *Gromphadorhina portentosa*. *Anim. Behav.* **1995**, *50*, 719–729. [CrossRef]
49. Clark, D.; Moore, A. Social Communication in the Madagascar Hissing Cockroach: Features of Male Courtship Hisses and a Comparison of Courtship and Agonistic Hisses. *Behaviour* **1995**, *132*, 401–417. [CrossRef]
50. Mack, C. *Winner and Loser Effects in Madagascar Hissing Cockroaches (Gromphadorhina portentosa)*; Bucknell University: Lewisburg, PA, USA, 2022.
51. Brouwers, S.; Duchateau, M.J. Feasibility and Validity of the Animal Welfare Assessment Grid to Monitor the Welfare of Zoo-Housed Gorillas Gorilla Gorilla Gorilla. *J. Zoo Aquar. Res.* **2021**, *9*, 208–217. [CrossRef]
52. Analog Devices Inc. *Thermochron iButton*; Analog Devices Inc.: Wilmington, MA, USA, 2015.
53. Mulder, P.; Shufran, A. *Madagascar Hissing Cockroaches: Information and Care*; Oklahoma State University: Stillwater, OK, USA, 2016.
54. R Core Team. *R: A Language and Environment for Statistical Computing*; R Foundation for Statistical Computing: Vienna, Austria, 2020.
55. Bradt, D.L.; Hoback, W.W.; Kard, B.M. American Cockroach Response to Cold Temperatures. *Southwest. Entomol.* **2018**, *43*, 335–342. [CrossRef]
56. Halsey, L.G.; Matthews, P.G.D.; Rezende, E.L.; Chauvaud, L.; Robson, A.A. The interactions between temperature and activity levels in driving metabolic rate: Theory, with empirical validation from contrasting ectotherms. *Oecologia* **2015**, *177*, 1117–1129. [CrossRef] [PubMed]

57. Arrant, A.E.; Schramm-Sapyta, N.L.; Kuhn, C.M. Use of the Light/Dark Test for Anxiety in Adult and Adolescent Male Rats. *Behav. Brain Res.* **2013**, *256*, 119–127. [CrossRef] [PubMed]
58. Zhukovskaya, M.; Novikova, E.; Saari, P.; Frolov, R.V. Behavioral Responses to Visual Overstimulation in the Cockroach *Periplaneta americana* L. *J. Comp. Physiol. A* **2017**, *203*, 1007–1015. [CrossRef]
59. Zewde, A.M.; Yu, F.; Nayak, S.; Tallarida, C.; Reitz, A.B.; Kirby, L.G.; Rawls, S.M. PLDT (Planarian Light/Dark Test): An Invertebrate Assay to Quantify Defensive Responding and Study Anxiety-like Effects. *J. Neurosci. Methods* **2018**, *293*, 284–288. [CrossRef] [PubMed]
60. Laurent Salazar, M.-O.; Planas-Sitjà, I.; Sempo, G.; Deneubourg, J.-L. Individual Thigmotactic Preference Affects the Fleeing Behavior of the American Cockroach (Blattodea: Blattidae). *J. Insect Sci.* **2018**, *18*, 9. [CrossRef]
61. Chen, Y.-R.; Li, D.-W.; Wang, H.-P.; Lin, S.-S.; Yang, E.-C. The Impact of Thigmotaxis Deprivation on the Development of the German Cockroach (*Blattella germanica*). *iScience* **2022**, *25*, 104802. [CrossRef]
62. Yoder, K. Food Resource Based Agonistic Interactions in Madagascar Hissing Cockroaches (*Gromphadorhina portentosa*). 2014. Available online: https://www.researchgate.net/publication/269410065_Food_Resource_Based_Agonistic_Interactions_in_Madagascar_Hissing_Cockroaches_Gromphadorhina_portentosa#fullTextFileContent (accessed on 29 August 2023).
63. Varadínová, Z.; Stejskal, V.; Frynta, D. Patterns of Aggregation Behaviour in Six Species of Cockroach: Comparing Two Experimental Approaches: Aggregation Behaviour of Cockroaches. *Entomol. Exp. Appl.* **2010**, *136*, 184–190. [CrossRef]
64. Sherwen, S.L.; Hemsworth, P.H. The Visitor Effect on Zoo Animals: Implications and Opportunities for Zoo Animal Welfare. *Animals* **2019**, *9*, 366. [CrossRef]
65. van Casteren, A.; Codd, J.R. Foot Morphology and Substrate Adhesion in the Madagascan Hissing Cockroach, *Gromphadorhina portentosa*. *J. Insect Sci.* **2010**, *10*, 40. [CrossRef]
66. Clark, D.; Shanklin, D. *Madagascar Hissing Cockroaches (Gromphadorhina portentosa)*; University of Kentucky: Lexington, KY, USA, 1995.
67. Broom, D.M. Indicators of Poor Welfare. *Br. Vet. J.* **1986**, *142*, 524–526. [CrossRef]
68. Hill, S.P.; Broom, D.M. Measuring Zoo Animal Welfare: Theory and Practice. *Zoo Biol.* **2009**, *28*, 531–544. [CrossRef] [PubMed]
69. Manteca, X.; Amat, M.; Salas, M.; Temple, D. Animal-Based Indicators to Assess Welfare in Zoo Animals. *CABI Rev.* **2016**, *2016*, 1–10. [CrossRef]
70. Veasey, J.S.; Waran, N.K.; Young, R.J. On Comparing the Behaviour of Zoo Housed Animals with Wild Conspecifics as a Welfare Indicator. *Anim. Welf.* **1996**, *5*, 13–24. [CrossRef]
71. Howell, C.P.; Cheyne, S.M. Complexities of Using Wild versus Captive Activity Budget Comparisons for Assessing Captive Primate Welfare. *J. Appl. Anim. Welf. Sci.* **2019**, *22*, 78–96. [CrossRef]

animals

MDPI

Review

Increasing Risks to the Health of the Invertebrates—Balancing between Harm and Benefit

Tatiana V. Kuznetsova [1,*], Valentina A. Kudryavtseva [1] and Larisa L. Kapranova [2]

[1] St. Petersburg Federal Research Center of the Russian Academy of Sciences, 199178 St. Petersburg, Russia; valenkud@yandex.ru

[2] A.O. Kovalevsky Institute of Biology of the Southern Seas of the Russian Academy of Sciences, 299011 Sevastopol, Russia; lar_sa1980@mail.ru

* Correspondence: kuznetsova_tv@bk.ru; Tel.: +7-950-044-38-64

Simple Summary: Emerging challenges associated with the COVID-19 (SARS-CoV-2) pandemic include the unpredictable biological effects of highly recommended detergents and disinfectants to reduce the risk of human infection. How these classes of substances affect living organisms, especially the health of the invertebrates as common representatives of biota, is still unclear. Due to the exceptional ability of disinfectants and detergents to penetrate biological membranes and alter the pH of the environment, these classes of substances, together with other environmental pollutants of concern (e.g., heavy metals), can have pronounced negative effects on macrobenthic invertebrates. The review analyzes the current state of research in this area and discusses ways to reduce risks to invertebrates.

Abstract: The article discusses the issue of extensive use of detergents and sanitizers in the time of new challenges associated with the COVID-19 (SARS-CoV-2) pandemic. These agents could pose threats to the existence of free-living invertebrates as essential components of the ecosystem. The biological effects of the mentioned classes of substances, their metabolites, and combined effects in the mixture have not been studied enough. The main challenges in trying to balance the threats and benefits of using such substances are the lack of knowledge of the biological effects of these products, the gaps in testing invertebrates' responses, and changes in environment-related regulations to minimize risks to animals and humans. Numerous studies in this field still leave research gaps, particularly concerning the combined toxicity of well-known and widely used disinfectants, surfactants, and heavy metals, posing potential future challenges. Additionally, the review identified the need for additional testing of invertebrates for their sensitivity to disinfectants and surfactants of different compositions, including improved (non-invasive) methods, studies for early life stages, and comparative studies of species resilience.

Keywords: surfactants; sanitizers; heavy metals; biological effects; legislation and nature protection; non-invasive monitoring methods; invertebrate welfare

1. Introduction

The SARS-CoV-2 virus spread has brought new challenges not only to the current epidemiological situation, but also raised fears of threats to environmental safety [1,2]. Since the first months of recording the pandemic, the World Health Organization (WHO) and the Russian Federal Service for Supervision of Consumer Rights and Human Welfare Protection (Rospotrebnadzor) have recommended the use of disinfectants and sanitizers to prevent infection and spread of the disease in crowded places, including hospitals, transport, institutions, schools, and households [3,4].

The purpose of these agents is to combat microorganisms and viruses. The majority of products that enter the distribution network include ethyl or isopropyl alcohol, propylene glycol, triclosan, formic, and, occasionally, salicylic acid, in addition to various fragrances and other ingredients.

However, people who use sanitizers frequently complain about dry skin, peeling, occasional skin flushing, and shortness of breath after several days of use, which manifest the adverse effects of these agents on the human body.

While a large proportion of the substances used in detergent formulations are subject to compulsory registration within the regulation of the Registration, Evaluation, Authorisation, and Restriction of Chemicals (REACH), there is still often little or even nothing known about their environmental impact. In only a few cases they are included in water monitoring programs (e.g., benzotriazole).

Wastewater from hospitals may contain a mixture of different drugs, radionuclides, anionic, cationic and amphoteric/non-ionic surfactants, disinfectants, and pathogenic microorganisms. The problem with most drugs is that, even after ordinary treatment, they are still present in wastewater [5], violating water quality requirements prescribed in Directive 98/83/EC [6].

All of these substances and their metabolites are released to the environment, where they may further react and interact with other materials, pollutants, or their metabolites. Hospital wastewater, moreover, contributes to the high resistance of bacteria to antibiotics. This subsequently affects the long-term ecological balance under chronic exposure [7]. Wastewater treatment plants are typically not designed to remove these chemicals both from biologically treated wastewater and from sludge. Sanitizers and disinfectants are mostly concentrated in the sludge from these plants. Common pathways of the surfactant and disinfectant entry into the environment and the main targets are presented in the diagram (Figure 1).

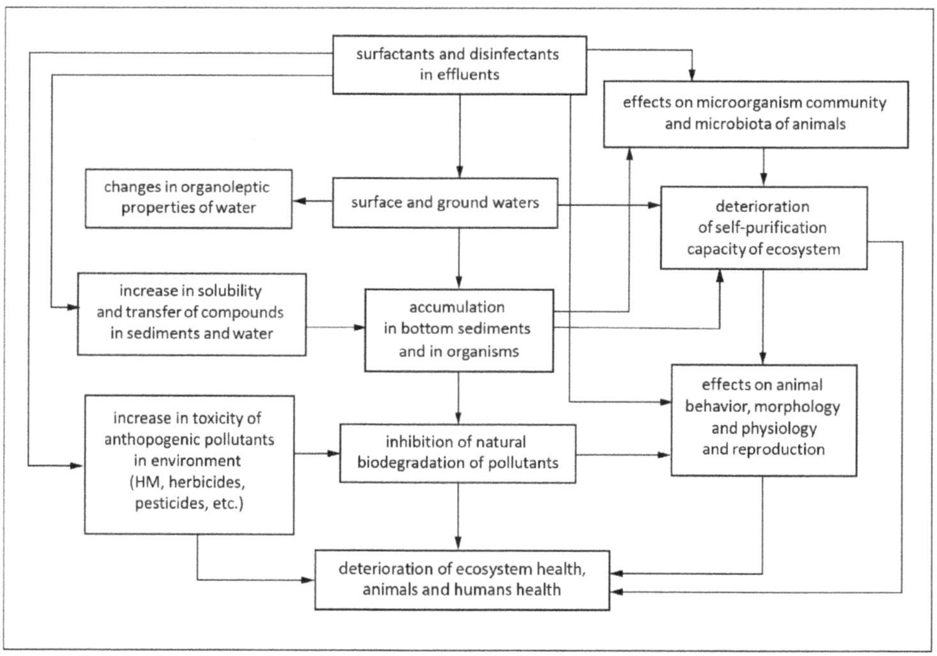

Figure 1. Common pathways of surfactant and disinfectant entry into the environment and the main consequences.

Using simple chemical analysis, it is not possible to assess the complex interactions and synergy of these substances and their actual effects on biological systems, as well as the threats to the integrity of the ecosystem under impact, including its abiotic and biotic components. In these cases, biological tests provide reliable information about the effects of the mentioned materials on the quality of treated wastewater entering natural water bodies and their effects on ecosystems and human health.

At present, there are no particular technologies and techniques for removing these wastewater contaminants and their metabolites. As a result, the levels of disinfectants in wastewater, and subsequently in surface waters, including those used by fisheries, will rapidly increase.

To a larger extent, the accumulation of sanitizers and their metabolites in aquatic environments can disturb the safety of both terrestrial and aquatic flora and fauna. This situation undoubtedly raises concerns among environmental scientists, physicians, environmental agency experts, and the general public.

Thus, there is a pressing need for novel information to evaluate the possible hazards of negative impacts, which can often be delayed, on the health of animals and humans. This evaluation should take into consideration the potential combined effects of contaminants in complex aquatic systems like surface waters.

2. General Information on Disinfectants: Benefits and Harm

In all cases, US EPA (2023) recommends the use of surfactants of different origins and compositions before disinfection [5].

Disinfectants are defined as "antimicrobial and antiviral products" [8]. In different countries, Environmental Protection Agencies (EPA) regulate and register antimicrobial agents and ensure that these agents meet minimum requirements for the registration to certify that they are effective against certain types of microorganisms or viruses. The use of disinfectants (including sanitizers) and antiseptics, recommended by WHO, is projected to increase [1]. However, it still requires consideration of the balance of benefits (killing of dangerous viruses and microorganisms) and possible harmful effects (indirect environmental and health impacts).

On the other hand, disinfectants have been often and successfully used in agriculture and aquaculture. In aquaculture, they help fight not only microorganisms and viruses, but can also significantly reduce the number of parasites and reduce the number of fungal diseases in farmed fish, crustaceans, and mollusks (e.g., [7]).

The most common disinfectants are chlorine products in their various formulations and forms (solution, solid, or powdery material).

At the same time, elevated levels of chlorine have the potential to cause skin or mucosal irritation, along with the chlorine odor side effects that may affect susceptible individuals like those with asthma. Thus, the balance between advantages and disadvantages arising from the application of various types of disinfectants (e.g., sanitizers) is currently a subject of active debate among scientists and clinicians [7–10].

In Europe and North America, different standards of chlorine-containing products have been adopted. The concentrations of chlorine in commercially available products vary between 4% and 6%, with a suggested concentration of 0.1% sodium hypochlorite for non-healthcare purposes [9,10]. An alternative option is the utilization of 70–90% ethyl alcohol for surface disinfection.

Furthermore, the current situation requires the extensive utilization of household antiseptics such as hand sanitizers for skin hygiene [4]. Hand sanitizers can have the same ingredients as healthcare-grade antiseptics, or they may vary by including additional substances for hydrating and nourishing the skin, flagrant liquids, food dyes (as the safest ones), and other materials. The components that commonly make up hand sanitizers include monohydric alcohols, propylene glycol, amine-containing materials, benzalkonium chloride, and skin care products.

These products are typically categorized as cosmetics, allowing manufacturers to circumvent the need for efficacy testing of their formulations. Currently, the Eurasian Economic Union (EAEU) is developing a project, in which the use of antiseptics in cosmetics is proposed to be limited. However, the regulations in the field of environment protection are becoming stricter all over the world (e.g., [11]), which is an important step towards a significant improvement in the welfare of invertebrates and vertebrates, including humans.

The legislation of the Russian Federation [12] and the European Council Directive 2006/44/EC [13] establishes restrictions for different categories of water use and for water bodies of fishery importance, but the maximum allowable concentrations (MACs) of chemicals in Russia, in contrast with the European Union (EU) regulations, are the same for all categories of water bodies. Some MACs for fishery water bodies in Russia [12] are much stricter. For example, the MAC for Cu established in Russian regulations is significantly lower than the guideline values recommended by the EU Council [13]. The same case is with national environmental quality standards on surfactants, in which MACs can vary from 0.1 to 0.5–0.6 mg/L.

In Russia, environmental regulations do not contain standards or sanitary and hygienic requirements for bottom sediments. Analysis of scientific and methodological literature has shown that there are no unified methods for assessing the state and degree of transformation of technogenically altered soils and sediments, including those under the impact of detergents from herbicide and pesticide formulations.

Thus, the current widespread use (both in the number of persons involved and in amounts) of disinfectants may lead in the future to the uncontrolled release of components into the environment that can negatively affect living organisms and ecosystems.

A number of studies performed on invertebrates showed the impact of surfactants on their different physiological functions and morphological structures e.g., [14–17]. Surfactant toxicity seems to be mainly related to the alteration of the cellular ionic balance due to changes in the permeability of the cell membrane.

In invertebrates, the test for acute toxicity on the freshwater crustacean *Daphnia magna* Straus, 1820, is highly recommended according to the standard EN ISO 6341 [18]. The essence of the method is finding the concentration of the test substance that causes immobilization of 50% of *Daphnia magna* newborns. This method could be also applicable in the case of studying the effects of surfactants and sanitizers.

The impact of surfactants in combination with other pollutants, such as heavy metals, has not been extensively studied in surface waters. Heavy metals (HMs), such as copper, zinc, lead, cadmium, mercury, arsenic, and more, are high-priority environmental pollutants that demonstrate substantial bioavailability to living organisms. Investigating the factors that influence the bioavailability and penetration of these elements into organisms, as well as the mechanisms and pathways for their excretion from hydrobionts, is an important fundamental objective in the fields of aquatic ecotoxicology and environmental safety [19].

All the factors mentioned above may have a negative impact on the environmental quality and biota as a sensitive component of ecosystems. And the most threatened in such conditions are aquatic invertebrates.

3. Use of the Invertebrates in Biomonitoring Studies

Invertebrates are most often and commonly used in biomonitoring studies: about two-thirds of the methods for assessing the quality of flowing waters are based on the use of macrobenthic invertebrates. Among them, mollusks and crustaceans are in priority in biomonitoring [20–23], as well as in ecotoxicological experiments [24–26]. There are several reasons for choosing these macroinvertebrates as bioindicators of environmental quality [27–30].

Taking into account all the advantages of using these animals, attention should be paid to the fact that ecotoxicological studies and biomonitoring used as routine methods allowing identification of environmental threats often cause "cruel deaths of up to millions of aquatic animals every year" (e.g., [31]). And the data provided by the author of the cited report concern only estimates in relation to Poland.

Invertebrates are "living monitors" of our present and future. It should not be forgotten that due to the fact that mollusks, e.g., from the family Unionidae, are active filter-feeders (one individual filters 20–40 L per day), they maintain the health of the ecosystem through bioaccumulation and biodeposition, participate in the processes of mineralization and, thus, are actively involved in self-purification of water bodies where they live [32]. We agree with J. Mather (2023) that our anthropocentric approach caused "three problems for invertebrates: a lack of knowledge, negative attitudes, and a misunderstanding of their cognitive abilities, all leading to problems for welfare consideration" [33]. Unfortunately, invertebrates are still considered just as convenient and low-cost model objects in ecotoxicology and toxicity research [34].

Consequently, the research is centered on the capacity of mollusks and crustaceans to act as indicators of pollution in coastal waters resulting from the discharge of domestic wastewater containing sanitizers, surfactants, their metabolites, heavy metals, or combinations thereof. At the moment, only a few studies are available on this issue; some of them are discussed below.

4. Use of the Invertebrates as Biosensors in Water Quality Monitoring Systems and as Test Organisms in Studying Disinfectants and Surfactants Actions

Invertebrates, mainly mollusks and crustaceans, are successfully used as live monitors of natural surface or drinking water pollution. Automated systems for these purposes have long been actively used in Europe and the United States to monitor water quality, and are called Biological Early Warning Systems (BEWS) [8]. Mussels change their locomotor behavior in response to disturbing agents (e.g., toxins) by closing their shells in order to reduce exposure. Therefore, the frequency of valve opening and closing can be monitored to indicate the avoidance behavior [8]. The commercially available MosselMonitor® system has been used with five different bivalve species (three for freshwater and two for marine water).

In the mentioned BEWS, the heart rate (HR) of mollusks or crustaceans is used as an indicative parameter of current changes in flowing water quality. HR is an integrative measure of the physiological status of the organism, and heart rate variability can serve as a basis for early diagnosis of deterioration at the whole organism level.

Among aquatic invertebrates, mollusks are the most accurate analogs to mammals in terms of the general structure, functioning, and regulatory circuits of cardiac activity. The main parameters of heart rate variability of mollusks are similar to the corresponding values in human rhythmograms [35]. However, studies on the cardiac activity of mollusks and crayfish are quite rare, especially when using automated systems for non-invasive HR monitoring [22,26,36–42]. Literature on automated systems using mollusks as biosensors is summarized in the review [43].

In early studies, it was shown that crustaceans change rhythms of respiratory and cardiac activities in the presence of HMs [44–47]. Similar studies were later conducted on crayfish to assess the effects of water disinfection in aquaculture, for example, during its chlorination [48] or chloramination [41]. For example, a 10 mg/L solution of the biocide chloramine-T is commonly used in industry and aquaculture. In experiments on the crayfish *Astacus leptodactylus* (Esch., 1823), the 1-h exposure did not adversely affect the crayfish at chloramine-T concentration of 10 or 50 mg/L [41]. Only after the 1-day exposure at a concentration of 50 mg/L, an obvious toxic effect was noted, leading to significant energy loss in the crayfish.

In the crayfish *Pacifastacus leniusculus*, the effects of ClO_2 at various concentrations, ranging from 0.01 to 0.29 mg/L, on cardiac activity were studied [48]. The authors noted that in 32% of crayfish individuals, the circadian rhythmicity in the HR is disrupted even at low concentrations of ClO_2 (0.01–0.2 mg/L), while higher concentrations lead to random periods of tachycardia and bradycardia in all experimental individuals. Crayfish mortality was observed after 3 to 4 weeks of exposure at concentrations exceeding 0.2 mg/L of ClO_2. The results indicate the importance of maintaining a certain standard of ClO_2 levels during water treatment, in order to prevent chlorite levels exceeding the WHO guideline value [49].

These studies can serve as a basis for further research on the effects of sanitizers on functional indicators of crustaceans and mollusks. Despite the great interest in the fate of surfactants and disinfectants in freshwater, the effects of environmentally relevant concentrations on target or non-target invertebrates are still unclear.

It is known that a biocenosis responds to significant changes in environmental quality by changing the metabolic rate in the organisms inhabiting it. The efficiency of aerobic energy exchange in organisms, assessed by the rate of oxygen consumption, can serve as an indicator of aquatic environment quality [44–46]. Another indicative physiological parameter is the intensity of the filtering activity of mollusks [50,51]. The advantages of using these particular functional indicators, whose changes are linked to the organism's efforts to prevent negative impact, center around their ability to identify the first signs of pollutants affecting a living organism and to show an early decline in animal health.

It has been noted that synthetic surfactants and detergents have adverse effects on the physiological condition of organisms, the water quality for the hydrobiont welfare, and the self-purification potential of water bodies [32]. The assessment of water pollution is rendered more complex due to the existence of chemical and biological degradation products that are more harmful than the initial compounds [46,47].

Ryabuhina and co-authors [52] reported changes in the dynamics of survival of *Ceriodaphnia* spp. in water samples with detergent. Experiments revealed an increase in the toxicity of sodium dodecyl sulfate (SDS) in a concentration of 25 mg/L after the 15th day of exposure.

There are only a few experimental studies in Russia that have shown the effects of anionic and cationic surfactants, tetradecyl trimethyl ammonium bromide (TDTMA), and SDS, at different concentrations on the antioxidant system (AOS), valve gape, and cardiac activity of the Black Sea mussels (*Mytilus galloprovincialis* Lam.) [53–56]. Under the action of a cationic detergent TDTMA at a concentration of 0.8 mg/L (a value close to being environmentally relevant) for 8 days, significant changes in the AOS were detected. Marked alterations were observed in the peripheral tissues of mussels, specifically the gills and foot, that were directly exposed to TDTMA [53]. A six-fold increase in the activity of superoxide dismutase (SOD) responsible for O_2^- neutralization was observed in the gills. Concurrently, there was an increase in catalase (CAT) levels by 1.7 and 3.2 times in the gills and foot, respectively. The response of AOS to this surfactant demonstrated tissue specificity, with the hepatopancreas displaying the least sensitivity to the detergent, while the gills exhibited the highest sensitivity to such exposure. The role of AOS in the response of entire organisms to environmental challenges is discussed in [57,58].

Ostroumov (2001) showed that the increase in the SDS concentration in water up to 1.7 mg/L leads to changes in the behavior of the exposed *Mytilus galloprovincialis* with long periods of the mussel being with valves closed [50]. Under these conditions, mussels switch to anaerobic metabolism, and during extended periods of exposure, a condition of oxygen deprivation known as hypoxia may occur. The closure of the mussel's shell serves as an indication of the adverse impact of the surfactant solution on the mussel [55]. Nonetheless, when exposed to small amounts (0.3–0.5 mg/L) of SDS, mussels detect the presence of this substance in the water, which brings about subsequent alterations in their heart rate and valve movements over time. This underscores the importance of considering the adverse effects associated with low levels of surfactants, which lead to well-expressed disturbances in circadian rhythmicity, with the loss of the predominant activity of mussels at night [56].

Changes in reactive oxidative stress markers were shown by Messina et al. [17] under the action of similar concentrations of SDS (from 0.1 to 1 mg/L), but after a long-term exposure (18 days), in the same mollusk species. Two enzymes, CAT and SOD, involved in reactive oxidative stress have high activity in the digestive gland of the exposed mussels.

Another difficult problem in studying the effects of surfactants and disinfectants is determining the threshold concentrations of the above-mentioned substances. In studies conducted with the use of the SDS, it was shown for the same mollusk species (*Mytilus galloprovincialis* Lam.) from the same water area (Crimea, Sevastopol), that the identified threshold concentration (lowest observed effect concentration—LOEC) can vary significantly, with a difference of up to several times. For example, when studying the inhibition of feeding behavior in the presence of SDS using the photometric method, this value was determined to be 1.7 mg/L [50], and when using non-invasive cardiac activity monitoring it was 0.3–0.5 mg/L [56]. These results raise questions about variables in effective doses that affect the activity patterns of mussels. Therefore, it is highly noteworthy that these levels depend strongly on the sensitivity of the techniques used to measure effects, method of registration, laboratory maintenance of the mussels, and functional system of mussels being tested.

In the most general approximation, the diagram in Figure 2 shows the entry of the mentioned substances into ecosystems and the impact on invertebrates and vertebrates, including humans, as assessed by the multi-biomarker approach and bioassay methods.

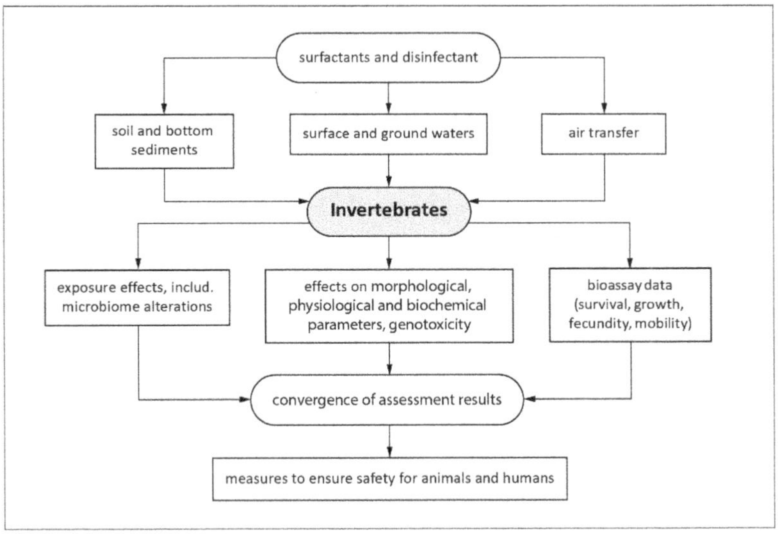

Figure 2. Scheme of the possible pathways of entry of surfactants and disinfectants into the environment and their impact on living organisms.

Surface waters are a usual object of environmental studies in terms of heavy metal concentrations, whose maximum allowable levels are set in regulations [6,11]. At the same time, under certain changes in the aquatic environment, HMs from sediments return to the water column via a number of physical, chemical, and biological processes. Thus, bottom sediments are a source of secondary pollution of water systems. Because the chemical behavior and environmental effects of metals in aquatic ecosystems are very complex, studies on HM behavior in sediments have become a hot topic in recent decades [59].

In terms of environmental pollution and its impact on living organisms, HMs in the environment are pollutants of high priority as announced in the Water Frame Directive [60]. The widespread distribution of HMs in the environment significantly endangers environmental safety. They spread far beyond the ranges of their initial emission and, unlike organic ecotoxicants, do not decompose but only change their speciation forms. The HM speciation forms are diverse and differ in toxicity. At the same time, "free" uncomplexed HM cations are considered to be the most toxic and the bound heavy metals are considered less dangerous because they lose bioavailability [61,62].

The main natural processes affecting the transformation and migration of HM compounds, including the most toxic labile forms, are the sorption of HM ions by dispersed minerals and complexation with humic acids. Therefore, to assess and predict the occurrence of environmental threats associated with the formation of labile HM forms in water bodies, it is necessary to study the basic regularities of metal ions binding to these materials. In this case, two main processes should be taken into account: the chemical equilibration of HM forms and the kinetics of speciation (sorption, complexation, etc.).

In a number of studies, the distribution of HMs, their accumulation in various ecosystems, in organs of living organisms, both aquatic and terrestrial, have been discussed [61–63]. However, it should be noted that in some species of crustaceans toxico-resistance to HM was observed [64]. One of the recently published studies is a valuable review based on observations in crustaceans [65]. It contains data on the elemental composition of HMs in the environment, their bioaccumulation in different animals' tissues, routes of excretion from the body, enzymatic responses, etc. In all models of biogeochemical cycles of pollutants that are created to assess hazards, it is necessary to take into account the interaction of pollutants with humic acids, since it radically changes both the chemical and toxicological behavior of harmful materials [66].

Another, no less important natural sorbent of HMs is finely dispersed minerals. A certain part of finely dispersed mineral suspensions consists of nanominerals, whose particles are of the nanometer size. Scientific studies raise concerns about the rapidly increasing production and consumption of nanomaterials, which will inevitably lead to their accumulation in the environment. The unique properties of nanoparticles, such as their huge surface area and abundance of surface reaction sites, facilitate their interaction with various environmental contaminants and toxicants, posing a new environmental threat [62,66].

The ability of sediments to accumulate ecotoxicants is influenced by physical, chemical, and biological factors. The physical factors include, first and foremost, the granulometric composition of sediments. Finely dispersed particles in sediments play a huge role in interfacial interactions, ion exchange, adsorption, structure formation, and immobilization of HMs; and studying the particle size distribution in relation to the sorption properties of sediments is of paramount importance [62,67–69].

At the same time, it is crucial to investigate the impact of biota and their metabolites on the fate of HMs in the environment [65]. It is worth mentioning that only labile hydrated ions or unstable complexes can easily cross cell membranes and are therefore biologically active. The most accessible forms of metals for benthic organisms are those dissolved in the pore water of sediments [70]. A study [71] explored how Cd^{2+} and Cu^{2+} ions affect the sorption of atrazine, a common herbicide, by bottom sediments. The research revealed that Cd enhances the sorption of atrazine synergistically, while Cu has an antagonistic effect. Therefore, understanding the patterns of the mutual influence of these toxicants on sorption processes seems necessary and very relevant.

Many studies have shown that heavy metal ions can accumulate in living organisms, disrupting their metabolic processes and inhibiting the production of proteins, e.g., enzymes [19,63,65]. Summarizing, we need to stress that all measurements, limitations, and regulations should aim at ensuring the health of the organisms and their environment. And, first of all, this concerns invertebrate animals, as sentinel species, and their welfare.

In assessing the biological impacts of heavy metal pollution on the environment, a common practice is to measure the bioconcentration factor (BCF) of heavy metals in animal tissues [72]. The accumulation of HMs in tissues, particularly for copper, zinc, lead, and cadmium, displays tissue specificity. Moreover, differential sensitivity to metals has been observed in mussel tissues. (e.g., [73–76]). BCF values of Cd, Cu, Pb, and Zn exceeding 1000 in mussels indicate that mussels have a great capacity to accumulate these metals. Thus, the mussels can transfer these metals to higher trophic levels, ultimately posing a potential hazard to human health.

In a number of studies, different biological effects of HMs on the physiological and biochemical indicators of aquatic organisms' state and mechanisms of organisms' adaptations were demonstrated [62,75,76]. etc. Most of these studies were carried out on bivalves and crustaceans, both of marine and freshwater origin, e.g., [64,65,75,76].

The ability of macrobenthic invertebrates, including mollusks and crustaceans, to absorb HMs depends on the metal speciation and the organism's characteristics. Therefore, it is essential to consider organisms' bioaccumulation capacity alongside data on metal concentrations and speciation forms in the abiotic parts of the ecosystem [77]. Voltammetry results indicate that the amounts of HMs, such as Cu, Cd, Zn, and Pb, that are readily bioavailable to these organisms can be affected by the pH of the aqueous solution used in the experiment. In the case of natural waters, the presence of disinfectants (sanitizers) and surfactants in water can change pH, thereby significantly affecting the solubility of other organic materials and increasing the number of labile forms of metals in the aquatic environment.

Summarizing, the accumulation of detergents, sanitizers, or their metabolites together with other environmental pollutants in the aquatic environment can disrupt the environmental safety/welfare of terrestrial and aquatic biota. Representatives of invertebrates play a key role in different ecosystems as environmental engineers (bioconstructors) [78]. In the presence of detergents that significantly change the pH of the environment, these animals may experience disturbances in the functioning of mucous structures and, consequently, excretory functions (for example, the production of secretions and feces). This disrupts building their tube-dwellings and producing similar materials for other animals in their common ecosystem [78]. Therefore, further research is needed on the effects of detergents and disinfectants on the essential functional systems of invertebrates.

5. Possible Ways to Reduce Risks for Animals and Humans

The current EU strategy on the protection of animals includes the implementation of the concept called "3Rs", i.e., reducing the number of animals in experiments (reduction), reducing the deprivation in the course of the experiments (refinement) and the ultimate aim is to completely replace testing on vertebrate animals (replacement) by alternative in vitro methods in the practice of research facilities. Yet, no specific invertebrate species are mentioned there.

One of the possible ways to reduce the intake of emerging substances toxic to animals (in particular, aquatic invertebrates) and humans may be seeking, replacing and further using disinfectants and surfactants that are less hazardous in their effects on biota. For example, according to the Centers for Disease Control and Prevention (CDC), vigorous handwashing in warm water with plain soap for at least 20 s is sufficient to fight germs in most cases. When plain soap and water are not available, the use of an alcohol-based hand sanitizer product is a better option than soap containing triclosan which is widely used as a disinfectant in households.

Another approach is the use of a variety of different test organisms from different trophic levels to ensure the negative or positive effects of disinfectants on living organisms. This approach is highly recommended by the Baltic Marine Environment Protection Commission (Helsinki Commission, HELCOM) for studying the toxicity of individual substances or their mixtures on benthic invertebrates [79].

It is necessary to note that the introduction of new advanced methods can significantly reduce freshwater invertebrate mortality and help in a more thorough assessment of the negative effects of toxicants (see SETAC Europe, 2022, discussion [80]). This concerns, for example, non-invasive (non-destructive) assessment methods, namely recording certain physiological functions (respiration, oxygen consumption, cardiac activity, and valve or gill movements) in mussels, clams, and crustaceans, where possible. Development of these assessment methods and techniques for their better performance can also become an important step towards a significant improvement in the welfare of aquatic invertebrates.

At present, the recommendations for disinfectant/detergent disposal routes differ according to the regionally established disposal structures. For example, according to Recommendations of the German Environment Agency disinfection of wastewater is accomplished both by filtering out harmful microorganisms and by adding disinfectant chemicals [80]. Water is disinfected so as to kill any pathogens that pass through filters and to provide a residual dose of disinfectant to kill or inactivate potentially harmful microorganisms in the storage and distribution systems.

In nowadays practice, chlorine dioxide ClO_2 is relatively rarely used in disinfection, because in some circumstances it may create excessive amounts of chlorite, whose release is strictly regulated, e.g., in the United States, to be at the lowest possible levels [11,50].

While a large proportion of the substances used in detergent formulations are subject to compulsory registration under the Registration, Evaluation, Authorization, and Restriction of Chemicals (REACH), there is still often too little or nothing known about their environmental properties. Accordingly, further investigations on levels of ingredients from detergent formulations in water are needed. First, the problematic ingredients relevant to monitoring should be identified. This requires, among other things, information on their toxicological relevance. In order to prioritize substances in the future in terms of monitoring and possible restrictions, the Germany Protection Agency—Umweltbundesamt (UBA)—is currently developing its own public information system, which will provide the necessary information about detergent ingredients by the end of 2024 at the latest [81].

Creating and improving evaluation bases and criteria for pollutants is of importance on a worldwide scale. Research into the entry of poorly biodegradable substances from detergents into waters remains one of the most important monitoring programs. There is a need for discussion on the restriction of poorly degradable substances and the development of analytical methods as a prerequisite for targeted monitoring.

One of the possible ways to reduce the content of disinfectants and detergents in the environment is phasing out the agricultural use of sewage sludge. The recommendations for sludge disposal routes differ according to the regionally established disposal structures [81].

Another demand is using potential synergies between the EU Council directives that provide measures to reduce the emission of disinfectants and detergents and their by-products in the environment.

Moreover, information on the effective use of disinfectants and surfactants, applied doses, and rules on disposal and storage of hazardous wastes should be disseminated among the population and attract social attention.

6. Conclusions

In the face of new challenges associated with the SARS-CoV-2 pandemic, greater attention should be paid to the compromise between harm and benefit in the application of disinfectants and sanitizers in our lives. It is necessary to take a closer look at the possible negative effects of these substances on the health of the Invertebrates already suffering from environmental pollution. Under development are new non-invasive methods of testing possible negative effects of these substances and their combined effects with high-priority pollutants (from the List of HELCOM) on organisms, especially aquatic invertebrates (as the main target species) of known contaminants or recently synthesized harmful substances, which could pose threats to animals' welfare.

Perhaps, there will be further proposals on this issue to ensure the welfare of the Invertebrates in light of the novel coronavirus pandemic and its associated challenges.

Author Contributions: Conception, writing—original draft preparation, T.V.K.; writing—review and editing—T.V.K., V.A.K. and L.L.K. All authors have read and agreed to the published version of the manuscript.

Funding: This research was carried out within the framework of SPC RAS "Scientific basis for assessing the health of ecosystems in North-West Russia and preventing threats to environmental safety" (FFZF-2022-0011), "Study of the transformation patterns of the cumulative technogenic background of natural and economic systems in the Gulf of Finland basin" (FFZF-2022-0014), and partly within the framework of state research assignment for IBSS RAS "Comprehensive study of the functioning mechanisms of marine biotechnological complexes with the aim of obtaining bioactive substances from hydrobionts" (No. 124022400152-1).

Acknowledgments: The authors thank William Winlow for his helpful advice on structuring this review. The authors are grateful to the anonymous reviewers for their valuable comments.

Conflicts of Interest: The authors declare no conflicts of interest.

References

1. *Report of the WHO-China Joint Mission on Coronavirus Disease 2019 (COVID-19)*; World Health Organization: Geneva, Switzerland, 2020. Available online: https://www.who.int/docs/default-source/coronaviruse/who-china-joint-mission-on-covid-19-final-report.pdf (accessed on 23 February 2023).
2. *Key Messages and Actions for COVID-19 Prevention and Control in Schools*; World Health Organization: Geneva, Switzerland, 2020. Available online: https://www.who.int/docs/defaultsource/coronaviruse/key-messages-and-actions-for-covid-19-prevention-and-controlin-schools-march-2020.pdf (accessed on 23 February 2023).
3. List N: Disinfectants for Use against SARS-CoV-2. US EPA. 2020. Available online: https://www.epa.gov/pesticide-registration/list-ndisinfectants-use-against-SARS-CoV-2 (accessed on 9 March 2023).
4. US EPA. Proposed Revisions to EPA's Safer Choice Standard. Docket ID EPA-HQ-OPPT-2023-0520. 2023. Available online: https://www.epa.gov/system/files/documents/2023-11/proposed-changes-to-epas-safer-choice-standard.pdf (accessed on 23 February 2023).
5. Escher, B.I.; Bramaz, N.; Ort, C. JEM spotlight: Monitoring the treatment efficiency of a full scale ozonation on a sewage treatment plant with a mode-of-action based test battery. *J. Environ. Monit.* **2009**, *11*, 1836–1846. [CrossRef]
6. EU Council Directive 98/83/EC from 3 November 1998. On the Quality of Water Intended for Human Consumption. Available online: http://data.europa.eu/eli/dir/1998/83/oj (accessed on 23 February 2023).
7. Jirova, G.; Wittlingerova, Z.; Zimova, M.; Vlkova, A.; Wittlerova, M.; Dvorakova, M.; Jirova, D. Bioindicators of wastewater ecotoxicity. *Neuro Endocrinol. Lett.* **2016**, *37* (Suppl. S1), 17–24. [PubMed]
8. US EPA. *Technologies and Techniques for Early Warning Systems to Monitor and Evaluate Drinking Water Quality. A State-of-Art Review*; Report EPA/600/R-05/156; US Environment Protection Agency Office of Water Office of Science and Technology, Health and Ecological Criteria: Wanshington, DC, USA, 2005; 236p. Available online: http://EPA%20report%202005.pdf (accessed on 26 May 2023).
9. Pereira, S.S.P.; Oliveira, H.M.; de Turrini, R.N.T.; Lacerda, R.A. Disinfection with sodium hypochlorite in hospital environmental surfaces in the reduction of contamination and infection prevention: A systematic review. *Rev. Esc. Enferm. USP* **2015**, *49*, 0681–0688. [CrossRef]
10. Köhler, A.T.; Rodloff, A.C.; Labahn, M.; Reinhardt, M.; Truyen, U.; Speck, S. Efficacy of sodium hypochlorite against multidrug-resistant Gram-negative bacteria. *J. Hosp. Infect.* **2018**, *100*, e40–e46. [CrossRef]
11. Directive 2008/105/EC of the European Parliament and of the Council of 16 December 2008 on environmental quality standards in the field of water policy, amending and subsequently repealing Council Directives 82/176/EEC, 83/513/EEC, 84/156/EEC, 84/491/EEC, 86/280/EEC and amending Directive 2000/60/EC. *Off. J.* **2008**, *L 348*, 84–97. Available online: http://eur-lex.europa.eu%E2%80%BAeli/dir/2008/105 (accessed on 25 February 2023).
12. The Order of the Federal Agency for Fisheries of 18.01.2010, № 20 «On approval of standards for water quality fishery water bodies, including the maximum permissible concentrations of harmful substances in the waters of fishery water bodies». *Rossijskaja gazeta*, 5 March 2010.
13. Directive 2006/44/EC of the European Parliament and of the Council of 6 September 2006 on the quality of fresh waters needing protection or improvement in order to support fish life. *Off. J. Eur. Union* **2006**, *L 264*, 20–35
14. Slye, J.L.; Kennedy, J.H.; Johnson, D.R.; Atkinson, S.F.; Dyer, S.D.; Ciarlo, M.; Stanton, K.; Sanderson, H.; Nielsen, A.M.; Price, B.B. Relationships between benthic macroinvertebrate community structure and geospatial habitat, in-stream water chemistry, and surfactants in the effluent-dominated Trinity River, Texas, USA. *Environ. Toxicol. Chem.* **2011**, *30*, 1127–1138. [CrossRef]
15. Gagné, F.; Chantale, A.; Fortier, M.; Fournier, M. Immunotoxic potential of aeration lagoon effluents for the treatment of domestic and hospital wastewaters in the freshwater mussel *Elliptio complanata*. *J. Environ. Sci.* **2012**, *24*, 781–789. [CrossRef]
16. Gilles, P. Cumulative impacts of urban runoff and municipal wastewater effluents on wild freshwater mussels (*Lasmigona costata*). *Sci. Total Environ.* **2012**, *431*, 348–356. [CrossRef]
17. Messina, C.M.; Faggio, C.; Laudicella, V.A.; Sanfilippo, M.; Trischitta, F.; Santulli, A. Effect of sodium dodecyl sulfate (SDS) on stress response in the Mediterranean mussel (*Mytilus galloprovincialis*): Regulatory volume decrease (RVD) and modulation of biochemical markers related to oxidative stress. *Aquat. Toxicol.* **2014**, *157*, 94–100. [CrossRef]
18. EN ISO 6341 Water Quality—Determination of the Inhibition of the Mobility of *Daphnia magna* Straus (Cladocera, Crustacea)—Acute Toxicity Test; ISO: Geneva, Switzerland, 1996.
19. Moiseenko, T.I. *Vodnaya Ekotoksikologiya: Teoreticheskie i Prikladnye Aspekty*; Nauka: Moscow, Russia, 2009; 399p. (In Russian)
20. Salánki, J.; Farkas, A.; Kamardina, T.; Rózsa, K.S. Molluscs in biological monitoring of water quality. *Toxicol. Let.* **2003**, *140–141*, 403–410. [CrossRef]
21. Depledge, M.H.; Galloway, T.S. Healthy animals, healthy ecosystems. *Front. Ecol. Environment.* **2005**, *3*, 251–258. [CrossRef]
22. Kuklina, I.; Kouba, A.; Kozák, P. Real-time monitoring of water quality using fish and crayfish as bio-indicators: A review. *Environ. Monit. Assess.* **2013**, *185*, 5043–5053. [CrossRef]

23. Hook, S.E.; Gallagher, E.P.; Batley, G.E. The role of biomarkers in the assessment of aquatic ecosystem health. *Integr. Environ. Assess. Manag.* **2014**, *10*, 327–341. [CrossRef]
24. Handy, R.D.; Depledge, M.H. Physiological Responses: Their measurement and use as environmental biomarkers in ecotoxicology. *Ecotoxicology* **1999**, *8*, 329–349. [CrossRef]
25. Curtis, T.M.; Williamson, R.; Depledge, M.H. Simultaneous, long-term monitoring of valve and cardiac activity in the blue mussel *Mytilus edulis* exposed to copper. *Mar. Biol.* **2000**, *136*, 0837–0846. [CrossRef]
26. Kuznetsova, T.V.; Sladkova, S.V.; Kholodkevich, S.V. Evaluation of functional state of crayfish *Pontastacus leptodactylus* in normal and toxic environment by characteristics of their cardiac activity and hemolymph biochemical parameters. *J. Evol. Biochem. Physiol.* **2010**, *46*, 241–250. [CrossRef]
27. Widdows, J.; Donkin, P. Mussels and environmental contaminants: Bioaccumulation and physiological aspects. In *The Mussel Mytilus: Ecology, Physiology, Genetics and Aquaculture*; Gosling, E., Ed.; Elsevier: Amsterdam, The Netherlands, 1992; pp. 383–424.
28. Gruber, D.; Frago, C.H.; Rasnake, W.J. Automated biomonitors—First line of defence. *J. Aquat. Ecosyst. Health* **1994**, *3*, 87–92. [CrossRef]
29. Kramer, K.J.M.; Foekema, E.M. The "Musselmonitor®" as Biological Early Warning System. In *Biomonitors and Biomarkers as Indicators of Environmental Change 2*; Environmental Science Research; Butterworth, F.M., Gunatilaka, A., Gonsebatt, M.E., Eds.; Springer: Boston, MA, USA, 2001. [CrossRef]
30. Nikinmaa, M. Bioindicators and Biomarkers. In *An Introduction to Aquatic Toxicology*; Elsevier: Amsterdam, The Netherlands; Academic Press: Cambridge, MA, USA, 2014; Chapter 12; pp. 147–155. ISBN 978-012-411574-3. [CrossRef]
31. Koperski, P. Freshwater invertebrates—Neglected victims of biological monitoring: An Ethical View. *Ethics Environ.* **2022**, *27*, 29–57. [CrossRef]
32. Ostroumov, S.A. Biological mechanism of self-purification in natural water bodies and streams: Theory and applications. *Adv. Mod. Biol.* **2004**, *124*, 429–442.
33. Mather, J.A. Ethics and invertebrates: The problem is us. *Animals* **2023**, *13*, 2827. [CrossRef]
34. Curpan, A.S.; Impellitteri, F.; Plavan, G.; Ciobica, A.; Faggio, C. *Mytilus galloprovincialis*: An essential, low-cost model **organism** for the impact of xenobiotics on oxidative stress and public health. *Review. Comp. Biochem. Physiol. C Toxicol. Pharmacol.* **2022**, *256*, 109302. [CrossRef]
35. Bychkov, R.; Zhuravlev, V.; Kodirov, S.; Safonova, T. Cardiac inhibitory neurons in the snail *Achatina fulica*. *J. Brain Res.* **1997**, *38*, 263–278.
36. Depledge, M.H.; Aagard, A.; Györkös, P. Assessment of trace metal toxicity using molecular, physiological and behavioural biomarkers. *Mar. Pollut. Bull.* **1995**, *31*, 19–27. [CrossRef]
37. Depledge, M.H.; Lundeby, A.K.; Curtis, T.; Aagaar, A.; Andersen, B.B. Automated interpulse-duration assessment (AIDA): A new technique for detecting disturbances in cardiac activity in selected invertebrates. *Mar. Biol.* **1996**, *126*, 313–319. [CrossRef]
38. Pautsina, A.; Kuklina, I.; Stys, D.; Cisăr, P.; Kozak, P. Noninvasive crayfish cardiac activity monitoring system. *Limnol. Oceanogr. Meth.* **2014**, *12*, 670–679. [CrossRef]
39. Kholodkevich, S.V.; Kuznetsova, T.V.; Sharov, A.N.; Kurakin, A.S. Applicability of a bioelectronics cardiac monitoring system for the detection of biological effects of pollution in bioindicator species in the gulf of Finland. *J. Mar. Syst.* **2017**, *171*, 151–158. [CrossRef]
40. Kozak, P.; Policar, T.; Fedotov, V.P.; Kuznetsova, T.V.; Buřič, M.; Kholodkevich, S.V. Effect of chloride content in water on heart rate in narrow-clawed crayfish (*Astacus leptodactylus*). *Knowl. Manag. Aquatic Ecosyst.* **2009**, *394–395*, 1–10. [CrossRef]
41. Kuklina, I.; Sladkova, S.; Kouba, A.; Kholodkevich, S.; Kozák, P. Investigation of chloramine-T impact on crayfish *Astacus leptodactylus* (Esch., 1823) cardiac activity. *Environ. Sci. Pollut. Res.* **2014**, *21*, 10262–10269. [CrossRef]
42. Kholodkevich, S.V.; Kuznetsova, T.V.; Sladkova, S.V.; Kurakin, A.S.; Ivanov, A.V.; Lyubimtsev, V.A.; Kornienko, E.L.; Fedotov, V.P. Industrial Operation of the Biological Early Warning System BioArgus for Water Quality Control Using Crayfish as a Biosensor. In *Water Science and Sustainability*; Sustainable Development Goals Series; Pandey, B.W., Anand, S., Cham, P., Eds.; Springer: Cham, Switzerland, 2021. [CrossRef]
43. Vereycken, J.; Aldridge, D.S. Bivalve molluscs as biosensors of water quality: State water quality: State of the art and future directions. *Hydrobiologia* **2023**, *850*, 231–256. [CrossRef]
44. Spicer, J.I.; Weber, R.E. Respiratory impairment in crustaceans and molluscs due to exposure to heavy metals. *Comp. Biochem. Physiol.* **1991**, *100*, 339–342. [CrossRef]
45. Styrishave, B.; Rasmussen, A.D.; Depledge, M.H. The influence of bulk and trace metals on the circadian rhythm of heart rates in freshwater crayfish *Astacus astacus*. *Mar. Pollut. Bull.* **1995**, *31*, 87–92. [CrossRef]
46. Kolupaev, B.I. *Respiration of Hydrobionts in a Toxic Environment*; Publishing House of Kazan University: Kazan, Russia, 1992. (In Russian)
47. Martin, J.S.; Saker, M.L.; Teles, L.F.; Vasconcelos, V.M. Oxygen consumption by *Daphnia magna* Strauss as a marker of chemical stress in the aquatic environment. *Environ. Toxicol. Chem.* **2007**, *26*, 1987–1991. [CrossRef]
48. Malinovska, V.; Ložek, F.; Kuklina, I.; Císăr, P.; Kozák, P. Crayfish as bioindicators for monitoring ClO_2: A case study from a Brewery Water Treatment Facility. *Water* **2020**, *12*, 63. [CrossRef]
49. World Health Organization. Chlorine Dioxide, Chlorite and Chlorate in Drinking-water. In *Background Document for Development of WHO Guidelines for Drinking-Water Quality*; World Health Organization: Geneva, Switzerland, 2016; p. 24.

50. Ostroumov, S.A. An amphiphilic substance inhibits the mollusk capacity to filter out phytoplankton cells from water. *Biol. Bull.* **2001**, *28*, 95–102. [CrossRef]
51. Ostroumov, S.A.; Widdows, J. Inhibition of mussel suspension feeding by surfactants of three classes. *Hydrobiologia* **2006**, *556*, 381–386. [CrossRef]
52. Ryabuhina, E.V.; Botyazhova, O.A.; Nikiforova, J.A. Change of functional condition *Ceriodaphnia affinis* at influence of various factors of environment. In *Current Problems of Physiology and Biochemistry of Aquatic Organisms: Proceedings of the III International Conference and Young Scientists School (22–26 June 2010)*; Karelian Research Centre of RAS: Petrozavodsk, Russia, 2010; pp. 161–163.
53. Gostyukhina, O.L.; Soldatov, A.A.; Golovina, I.V. Influence of tetradecyltrimethyl ammonium bromide on the state of the enzymatic system of antioxidant defense of the tissues of the Black Sea mollusk *Mytilus galloprovincialis* Lam. *Rep. Natl. Acad Sci. Ukraine* **2007**, *11*, 147–151.
54. Trusevich, V.V.; Kuz'min, K.A.; Mishurov, V.J.; Zhuravskii, V.J.; Vyishkvarkova, E.V. Features of behavioral reactions of the Black Sea mussels *Mytilus galloprovicialis* in natural habitat. *Biol. Inland Waters* **2021**, *1*, 12–22. [CrossRef]
55. Trusevich, V.V.; Kuzmin, K.A.; Mishurov, V.J. Biomonitoring of the surface water quality with use of freshwater bivalvia molluscs. *Environ. Control Syst.* **2017**, *7*, 83–93. (In Russian)
56. Kuznetsova, T.; Kholodkevich, S. Comparative assessment of surface water quality through evaluation of physiological state of bioindicator species: Searching new biomarkers. In Proceedings of the 2015 4th Mediterranean Conference on Embedded Computing (MECO), Budva, Montenegro, 14–18 June 2015. [CrossRef]
57. Soldatov, A.A.; Gostyukhina, O.L.; Golovina, I.V. Functional states of antioxidant enzymatic complex of tissues of *Mytilus galloprovincialis* Lam. under conditions of oxidative stress. *J. Evol. Biochem. Physiol.* **2014**, *50*, 206–214. [CrossRef]
58. Chuiko, G.M. *Biomarkers in Hydroecotoxicology: Principles, Methods and Methodology, Practice of Use. Ecol. Monitoring. Part VIII. Current Probl. of Monitoring Freshwater Ecosystems: A Study Guide*; NNGU: Nizhnii Novgorod, Russia, 2014. (In Russian)
59. Rahman, Z.; Singh, V.P. The relative impact of toxic heavy metals (THMs) (arsenic (As), cadmium (Cd), chromium (Cr)(VI), mercury (Hg), and lead (Pb)) on the total environment: An overview. *Environ. Monit. Assess.* **2019**, *191*, 419. [CrossRef]
60. Water Frame Directive. Directive 2000/60/EC of the European Parliament and of the Council of 23 October 2000 establishing a framework for Community action in the field of water policy. *Off. J.* **2000**, *L 327*, 0001–0073. Available online: https://leap.unep.org/en (accessed on 23 February 2023).
61. Horowitz, A. *A Primer on Trace Metal-Sediment Chemistry*, 2nd revised ed.; Lewis Publishers: Chelsea, MI, USA, 1991; 136p.
62. Dash, S.; Borah, S.S.; Kalamdhad, A.S. Heavy metal pollution and potential ecological risk assessment for surficial sediments of Deepor Beel, India. *Ecol. Indic.* **2021**, *122*, 107265. [CrossRef]
63. DeForest, D.K.; Brix, K.V.; Adams, W.J. Assessing metal bioaccumulation in aquatic environments: The inverse relationship between bioaccumulation factors, trophic transfer factors and exposure concentration. *Aquat. Toxicol.* **2007**, *84*, 236–246. [CrossRef]
64. Strode, E.; Balode, M. Toxico-resistance of Baltic amphipod species to heavy metals. *Crustaceana* **2013**, *86*, 1007–1024. [CrossRef]
65. Banaee, M.; Zeidi, A.; Mikušková, N.; Faggio, K. Assessing metal toxicity on crustaceans in aquatic ecosystems: A comprehensive review. *Biol. Trace Elem. Res.* **2024**, *202*, 1–21. [CrossRef]
66. Nel, A.; Xia, T.; Madler, L.; Li, N. Toxic potential of materials at the nanolevel. *Science* **2006**, *311*, 622–627. [CrossRef]
67. Du Laing, G.; Rinklebe, J.; Vandecasteele, B.; Meers, E.; Tack, F.M.G. Trace metal behavior in estuarine and riverine floodplain soils and sediments: A review. *Sci. Total Environ.* **2009**, *407*, 3972–3985. [CrossRef]
68. Gadd, G.M. Microbial influence on metal mobility and application for bioremediation. *Geoderma* **2004**, *122*, 109–119. [CrossRef]
69. Kováčová, S.; Šturdík, E. Interactions between microorganisms and heavy metals including radionuclides. *Biol.-Sect. Cell. Mol. Biol.* **2002**, *57*, 651–663.
70. Levit, R.L.; Kudriavtseva, V.A.; Shigaeva, T.G. The effects of the main cations of natural aquatic media on zinc(II), cadmium(II), lead(II) and copper(II) sorption by aluminium oxide and kaolin. *Interdiscip. Sci. Appl. J. "Biosph."* **2014**, *6*, 382–387.
71. Qian, G.; Ao, W.; Yu, L. Combined effect of co-existing heavy metals and organophosphate pesticide on adsorption of atrazine to river sediments. *Korean J. Chem. Eng.* **2011**, *28*, 1200–1206. [CrossRef]
72. Mendoza-Carranza, M.; Sepulveda-Lozada, A.; Dias-Ferreira, C.; Geissen, V. Distribution and bioconcentration of heavy metals in a tropical aquatic food web: A case study of a tropical estuarine lagoon in SE Mexico. *Environ. Poll.* **2016**, *210*, 155–165. [CrossRef] [PubMed]
73. Brown, M.T.; Depledge, M.H. *Metabolism of Trace Metals in Aquatic Organisms*; Bebianno, M.J., Langston, W.J., Eds.; Chapman & Hall: London, UK, 1998.
74. Brown, R.J.; Galloway, T.S.; Lowe, D.; Browne, M.A.; Dissanayake, A.; Jones, M.B.; Depledge, M.H. Differential sensitivity of three marine invertebrates to copper assessed using multiple biomarkers. *Aquat. Toxicol.* **2004**, *66*, 267–278. [CrossRef] [PubMed]
75. Gundacker, C. Comparison of heavy metal bioaccumulation in freshwater molluscs of urban river habitats in Vienna. *Environ. Pollut.* **2000**, *110*, 61–71. [CrossRef] [PubMed]
76. Zarykhta, V.V.; Zhang, Z.; Kholodkevich, S.V.; Kuznetsova, T.V.; Sharov, A.N.; Zhang, Y.; Sun, K.; Lv, M.; Feng, Y. Comprehensive assessments of ecological states of Songhua River using chemical analysis and bivalves as bioindicators. *Environ. Sci. Pollut. Res.* **2019**, *26*, 33341–33350. [CrossRef] [PubMed]
77. Kudryavtseva, V.; Shigaeva, T.; Alekseeva, N. Heavy metals in the bottom sediments of the coastal zone of the eastern part of the Gulf of Finland. *E3S Web Conf.* **2021**, *265*, 02015. [CrossRef]

78. Lisco, S.N.; Acquafredda, P.; Gallicchio, S.; Sabato, L.; Bonifazi, A.; Cardone, F.; Corriero, G.; Gravina, M.F.; Pierri, C.; Moretti, M. The sedimentary dynamics of *Sabellaria alveolata* bioconstructions (Ostia, Tyrrhenian Sea, central Italy). *J. Palaeogeogr.* **2020**, *9*, 1–18. Available online: https://www.researchgate.net/publication/338548583 (accessed on 14 May 2024). [CrossRef]
79. HELCOM 2014. *BASE Project 2012–2014: Preparation of Biodiversity and Hazardous Substances Indicators with Targets That Reflect Good Environmental Status for HELCOM (Including the HELCOM CORESET Project) and Improvement of Russian Capacity to Participate in Operationalization of Those Indicators*; Baltic Marine Environment Protection Commission HELCOM: Helsinki, Finland, 2014.
80. SETAC Europe Annual Meeting, 2022, Copenhagen, Denmark, Discussion. Available online: https://www.oecd.org/chemicalsafety/risk-assessment/guiding-principles-and-key-elements-for-establishing-a-weight-of-evidence-for-chemical-assessment.pdf (accessed on 6 May 2022).
81. *Recommendations for Reducing Micropollutants in Wastewater*; German Environment Agency. Section II 2.1 General Aspects of Water and Soil; Helmecke, M. (Ed.) 2018; 60p, Available online: https://www.umweltbundesamt.de/sites/default/files/medien/1410/publikationen/180709_uba_pos_mikroverunreinigung_en_bf.pdf (accessed on 26 May 2023).

animals MDPI

Article

Effects of Artificial Light at Night on Fitness-Related Traits of Sea Urchin (*Heliocidaris crassispina*)

Xiuwen Xu, Zexianghua Wang, Xiuqi Jin, Keying Ding, Jingwen Yang and Tianming Wang *

National Engineering Laboratory of Marine Germplasm Resources Exploration and Utilization, Marine Science and Technology College, Zhejiang Ocean University, Zhoushan 316022, China; xiuwenxu1207@163.com (X.X.); wangzexianghua@zjou.edu.cn (Z.W.); 2021060@zjou.edu.cn (X.J.); dingkeying1228@163.com (K.D.); silence84309@163.com (J.Y.)

* Correspondence: wangtianming@zjou.edu.cn

Simple Summary: The sea urchin (*Heliocidaris crassispina*) is an ecologically important invertebrate in structuring marine benthic communities. Most dwell within intertidal regions, rendering them highly susceptible to elevated artificial light at night. However, their potential acclimation to artificial light exposure remains largely unexplored. This study investigates the changes of prolonged artificial light pollution on the fitness-related traits of sea urchins. Following a six-week exposure to artificial light at night, survival remained unaffected, while behavioral responses exhibited slower reactions. Concurrently, growth inhibition was observed in sea urchins, which might be attributed to reduced mouthparts weight and decreased food consumption. This study also revealed that sea urchin gonads are more susceptible to artificial light than guts. *Pax6* gene expression may serve as a sensitive indicator to assess the impact on the photosensitive system of sea urchins. These results increase our understanding of the effects of artificial light at night on sea urchins and provide valuable information about coastal animals' safety.

Abstract: Limited data are available regarding the effects of elevated coastal artificial light at night (ALAN) on intertidal echinoderms. In this study, we investigated the behavioral, morphological, and physiological responses of the sea urchin (*Heliocidaris crassispina*) after continuous exposure to ALAN at light intensities of 0.1, 300, and 600 Lux for 6 weeks. Our findings revealed that ALAN at 300 Lux substantially reduced food consumption, Lantern weight, and gonadosomatic index (GSI). On the other hand, ALAN at 600 Lux notably prolonged the righting and covering response times and elevated the 5-HIAA/5-HT ratio, while concurrently decreasing food consumption, body weight, Lantern weight, GSI, and *Pax6* gene expression. These results indicated that continuous exposure to ALAN could cause an adverse effect on fitness-related traits, including behavioral responses, growth, reproductive performance, and photoreception of sea urchins. The present study provides new insights on the impact of light pollution on echinoderms.

Keywords: ALAN; sea urchin; fitness; behavior; GSI; growth

1. Introduction

Artificial light at night (ALAN) changes the natural state of a dark environment from dark to bright by increasing the distribution and intensity of light in space [1–3]. As coastal population densities surge, the proliferation of ALAN is expanding at an annual growth rate of approximately 6% [2,4]. While ALAN offers substantial convenience for human activities, it concurrently appears to present a potential threat to coastal ecosystems [2,5–9]. For example, in shallow water coral reef ecosystems, recent studies reported the gametogenesis cycle of *Acropora millepora* was delayed or masked by exposure to ALAN [10]. In sandy beach ecosystems, sea turtle orientation was negatively affected by ALAN, impairing the ability of hatchlings to respond to natural orientation cues [11,12]. In seabird communities, increasing ALAN levels in and around nesting colonies could impact sea mew breeding behavior and, consequently, chick provisioning [13]. However, scant attention has been directed towards investigating the implications of ALAN in intertidal ecosystems, particularly concerning invertebrates such as slow-moving echinoderms. Therefore, a comprehensive grasp of how intertidal invertebrates' fitness reacts to ALAN exposure is imperative as it provides essential insights into potential ecological risks in coastal environments.

Sea urchin *Heliocidaris crassispina* (previously *Anthocidaris crassispina*) is an echinoderm species found across a wide range, from the intertidal zone to shallow waters in the Western Pacific region. Its habitat spans from the Japan Sea (at 38 °N) to Hainan Island (at 19 °N) [14–17]. This species holds a crucial role in intertidal ecosystems, acting both as an herbivorous grazer on macroalgal species such as *Ulva Lactuca* and as prey for crabs and starfish [18,19]. Echinoderm behaviors (such as righting, covering, and feeding) are important for their fitness but also affect intertidal ecosystems through effects on their predators, prey, and competitors [20]. Righting is the behavior of inverted sea urchins placed on their aboral surface to resume the posture with the aboral side up [21,22]. Covering behavior refers to sea urchins using their tubular feet and spines to move objects such as algae fragments, stones, and shells to the dorsal surface [23–25]. These two behaviors are used to help sea urchins defend against abiotic disturbances and predation [24]. Health-related attributes, such as food consumption, body weight, and gut-weight, alongside reproductive traits like gonad weight, constitute foundational elements influencing the fitness of sea urchins [20,21]. Serotonin serves as one of the monoamine neurotransmitters integral to regulating animal behavior [26]. It can be used to reflect the physiological state of sea urchins when their behavioral expression changes. *Pax6* is a pivotal opsin transcription factor gene within the animal photosensitive system. Deviations in *Pax6* gene expression could indicate a disruption of the light defense system in sea urchins [27].

The primary objective of this study is to examine the effects of prolonged artificial lighting on the righting and covering behaviors, health-related and reproductive traits, coelomic monoaminergic activity, and *Pax6* gene expression in *H. crassispina*. This investigation aims to offer novel insights into the repercussions of ALAN on intertidal creatures, alongside potential underlying mechanisms.

2. Materials and Methods

Before commencing the study, we meticulously reviewed animal experiments and welfare regulations, ensuring strict adherence. All procedures undertaken in this study received approval from the Institutional Animal Care and Use Committee of Zhejiang Ocean University.

2.1. Sea Urchin Materials

Sea urchins *H. crassispina* were collected from the waters near Zhoushan Islands (122°41′27.040″ E, 30°12′41.625″ N) by scuba diving and delivered to a 5000 L PVC tank in the Culture Enhancement Laboratory of Zhejiang Ocean University for acclimation. The PVC tank was supplied with filtered seawater at constant aeration and a flow rate of 10 L/min. The seawater was changed every 5 days and its temperature was set at 24.5–26.5 °C. Daytime lighting was provided by low-pressure mercury discharge fluorescent lamps (626–1023 Lux, 12 h light-12 h dark). Sea urchins were fed ad libitum with chopped *U. Lactuca* and their droppings were cleaned every day [21,28].

2.2. Experimental Design

After 3 weeks of acclimation, seventy-two sea urchins were selected from the PVC tank and weighed for this experiment (body weight ± SD =54.402 ± 6.240 g, n = 72). The weighed sea urchins were then randomly divided evenly into nine 150 L glass tanks (long × wide × deep = 60 cm × 50 cm × 50 cm) placed in the laboratory (n = 8 for each tank; three tanks for each group; n = 24 for each group). The glass tanks' condition during the experiment was similar to the PVC tank except for artificial light illumination at night. Nocturnal illumination stimulation was provided by metal halide streetlights (HC-ZY521, OSRAM, Munich, Germany) mounted on the glass tanks. Neutral density filters (248565, Grand Unified Optics, Beijing, China) were utilized to attenuate the intensity of the streetlights, thereby creating varying levels of nocturnal light intensities on the water surface, while ensuring no alteration to the light spectrum. Three groups were thus set up which, respectively, represent the simulated living environment in three situations (Figure 1a) [29–33]. The first group was the control group, simulating a moon-transparent night (ALAN = ~0.1 Lux) [29,33]. The second group was stimulated continuously with low artificial light to simulate the marine environment under the influence of city lights (ALAN = ~300 Lux) [29–31]. The third group was given continuous intense artificial light stimulation to simulate the marine environment of direct irradiation near industrial port facilities (ALAN = ~600 Lux) [32,33].

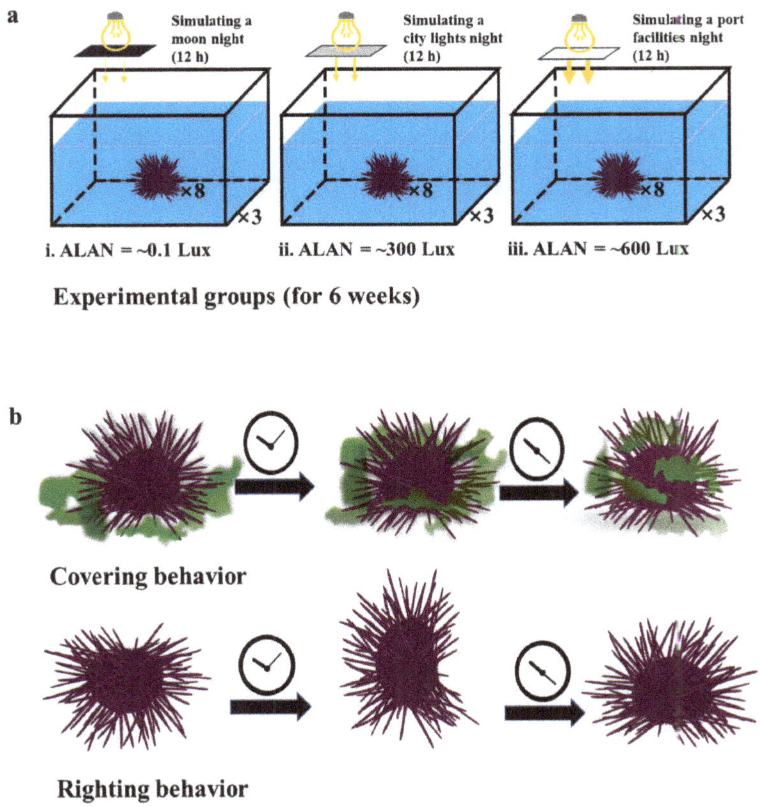

Figure 1. Schematic drawing of the three ALAN experimental groups (**a**) and the representative fitness behaviors (**b**) of sea urchins.

During the experiment, dead sea urchins (if any) were removed and recorded every day. Water quality assessments were performed at weekly intervals utilizing a portable water quality monitoring meter (YSI, Yellow Springs, OH, USA) (Supplementary Materials: Table S1). Six weeks later, the three groups of sea urchins that had been subjected to different long-term artificial light treatments were used for subsequent behavioral and physiological analysis.

2.3. Behavioral Analysis

Righting behavior, covering behavior, and food consumption of *H. crassispina* were measured in glass tanks (long × wide × deep = 40 × 30 × 30 cm; 20 cm of aeration water depth). Righting behavior [16,20–23]: A sea urchin was placed at the tank's center with its aboral side facing downward, and the time taken for it to reorient itself to the oral side facing downward was measured as the righting response time. If the sea urchin failed to reorient within 10 min, the righting response time was recorded as 600 s. This experimental protocol aligns with previous studies on covering behavior [23–25,27]: A sea urchin was carefully positioned at the central point of the tank, with *U. Lactuca* available ad libitum at the bottom. The time required for the sea urchin to cover its body with *U. Lactuca* was recorded as the covering response time. If the individual failed to engage in covering behavior within a 30-min timeframe, its covering response time was recorded as 1800 s (Figure 1b). Food consumption [16,20,21]: Four sea urchins were delicately placed at the center of the tank, with weighted *U. Lactuca* available ad libitum at the bottom. The sea urchins began to feed about two hours later. Food consumption was quantified by subtracting the remaining uneaten quantity of *U. Lactuca* in the tank after a 24-h period from the initially supplied amount. To mitigate observer bias, a double-blinded approach was employed for the recording and analysis of all behavioral data.

2.4. Sampling Protocol

The sampling procedure of *H. crassispina* was conducted by the method described in previous papers [34]. Test diameter, test height, and Aristotle's lantern length were assessed with a vernier caliper, with measurements recorded to the nearest 0.1 mm (CD-AX/C, MITUTOYO, Kawasaki city, Japan). Test weight, Aristotle's lantern weight, gonads weight, and gut-weight were measured to the nearest 0.01 g by an electronic balance (C3-6553, AS ONE, Osaka, Japan). Subsequently, coelomic fluid was extracted and subjected to centrifugation at 10,000 rpm for 10 min in a refrigerated centrifuge (4 °C). The resulting supernatant was meticulously collected and preserved at −80 °C for subsequent analysis of monoaminergic activity. Tubular feet were cut off and washed thoroughly with ice-cold artificial coelomic fluid (10 mM $CaCl_2$; 14 mM KCl; 50 mM $MgCl_2$; 398 mM NaCl; 1.7 mM Na_2HCO_3; 25 mM Na_2SO_4) [35], and frozen quickly in liquid nitrogen for gene expression analysis. Test height/test diameter [25] and gonads weight/test weight (GSI) [20] were subsequently calculated.

2.5. Physiological Analysis

The concentrations of 5-hydroxytryptamine (5-HT) and 5-hydroxyindoleacetic acid (5-HIAA) in body cavity liquid samples were measured by commercial enzyme-linked immunosorbent assay (ELISA) Assay Kits (Nanjing Jiancheng Bioengineering Institute, Nanjing, China) according to the manufacturer's instructions. These ELISA Assay Kits have been validated in marine life previously [36–39]. *Pax6* gene expression levels of tubular feet samples were measured using quantitative real-time polymerase chain reaction (qPCR) [40,41]. Total RNA was extracted using Trizol Reagent (Invitrogen, Carlsbad, CA, USA) containing bromochloropropane for phase separation. Afterward, RNA integrity was confirmed via electrophoresis on a 1.0% agarose gel, and quantification was conducted using a NanoDrop 2000 spectrophotometer (Thermo Scientific, Waltham, WA, USA). Subsequently, double-strand cDNA was synthesized from the total RNA template, and the construction of a cDNA library was achieved using the SMART cDNA construction kit (Clontech, Palo Alto, CA, USA). Primers for the study were designed utilizing Primer Premier 6 software (PREMIER Biosoft International, Palo Alto, CA, USA) and obtained commercially from Sangon Biotech (Shanghai, China). As a reference gene in sea urchins [42] (Table 1), the internal control gene *18S rRNA* was employed. Each qPCR reaction involved a 20 μL volume, including 10 μL of the 2× UltraSYBR Mixture, 0.2 μM specific forward and reverse primers, and 1.0 μL of the diluted cDNA template (10 ng/μL). The thermal cycling process was carried out on the ABI Prism 7900HT Sequence Detection System (PE Applied Bio-systems, Foster City, CA, USA), encompassing an initial denaturation step at 95 °C for 30 s, followed by 40 cycles of denaturation at 94 °C for 5 s, annealing at 60 °C for 30 s, and extension at 72 °C for 10 s. To confirm the specificity of amplification, a melting curve analysis of the PCR products was executed. Relative mRNA levels were determined using the $2^{-\Delta\Delta Ct}$ method [43]. The procedure of *Pax6* gene full-length conventional cloning of *H. crassispina* was fully described in the Supplementary Materials: *Procedure (Sp)*.

Table 1. Nucleotide sequences of the primers for qPCR.

Primer Names	Sequence (5′→3′)	Application	AT (°C)
Pax6-F	AAGGCTGAAGATGATGAAGA	qPCR of *Pax6* gene	58
Pax6-R	GGAATGATTGGAAGACTGAC		
18S-F	ACGAAGGAGAAGACAAGG	qPCR of *18S rRNA* gene	56
18S-R	AAGCCACAAACGACAGTA		

AT: Annealing temperature.

2.6. Data Statistics

Data throughout the text, tables, and figures are expressed as means ± standard error (S.E.). Statistical significance was established at $p < 0.05$. Data conforming to a normal distribution with equal variances underwent one-way analysis of variance (ANOVA) followed by Duncan's multiple range post hoc test for statistical examination. When necessary, data were subjected to appropriate transformations, such as square root, exponential, or logarithm, to meet the assumptions of normal distribution and homogeneity of variance. Statistical analyses were conducted utilizing SPSS 17.0 for Windows (SPSS, IBM, Armonk, NY, USA, 2007).

3. Results

3.1. Survival

We found that ALAN at 0.1, 300, and 600 Lux could not cause sea urchin death. There was no significant difference in the number of surviving sea urchins among the individuals exposed to 0.1, 300, and 600 Lux of ALAN. However, sea urchins exposed to ALAN at 600 Lux showed severe spine loss during the experiment.

3.2. Righting and Covering Behavior

We found that ALAN at 600 Lux significantly increased the righting response time of *H. crassispina* compared to those exposed to ALAN at 0.1 and 300 Lux ($p < 0.005$; $p < 0.005$). However, there was no significant difference between the individuals exposed to 0.1 and 300 Lux of ALAN ($p = 0.313$; Figure 2a). Similarly, ALAN at 600 Lux significantly increased the covering response time of *H. crassispina* compared to those exposed to ALAN at 0.1 and 300 Lux ($p < 0.005$; $p < 0.005$). However, there was no significant difference between the individuals exposed to 0.1 and 300 Lux of ALAN ($p = 0.442$; Figure 2b).

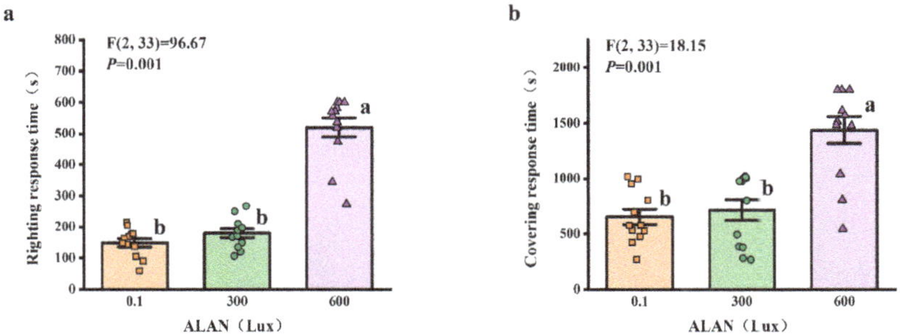

Figure 2. Effects of ALAN at different intensities for 6 weeks on righting response time (**a**) and covering response time (**b**) of sea urchins. Different letters indicate significant differences. Values are shown in data points and mean ± S.E. ($n = 12$).

3.3. Food Consumption

We found that the food consumption of *H. crassispina* significantly decreased after 6 weeks of ALAN exposure. Sea urchins exposed to ALAN at 300 Lux and 600 Lux resulted in a significant decrease in food consumption compared with the control (0.1 Lux) ($p < 0.05$; $p < 0.05$). However, there was no significant difference between the individuals exposed to 300 and 600 Lux of ALAN ($p = 0.558$; Figure 3).

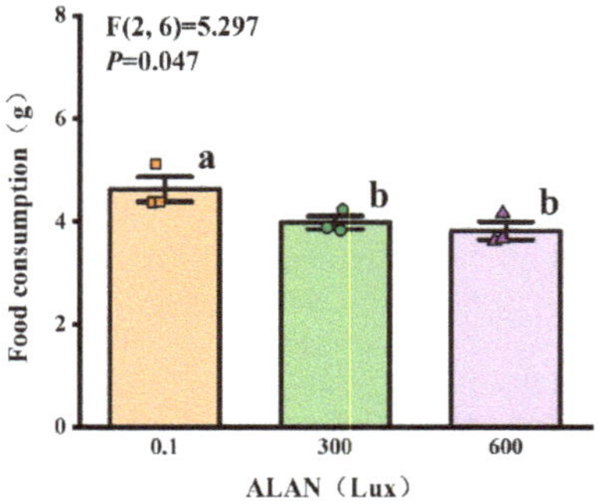

Figure 3. Effects of ALAN at different intensities for 6 weeks on food consumption of sea urchins. Different letters indicate significant differences. Values are shown in data points and mean ± S.E. ($n = 3$).

3.4. Body Size

We found that ALAN at 600 Lux significantly decreased the body weight of *H. crassispina* compared to those exposed to ALAN at 0.1 Lux ($p < 0.05$; Figure 4a). However, there was no significant difference between the individuals exposed to 0.1 and 300 Lux or 300 and 600 Lux of ALAN ($p = 0.133$; $p = 0.245$; Figure 4a). In addition, there was no significant difference in test height, test diameter, and test height/test diameter among the individuals exposed to 0.1, 300, and 600 Lux of ALAN (Figure 4b–d).

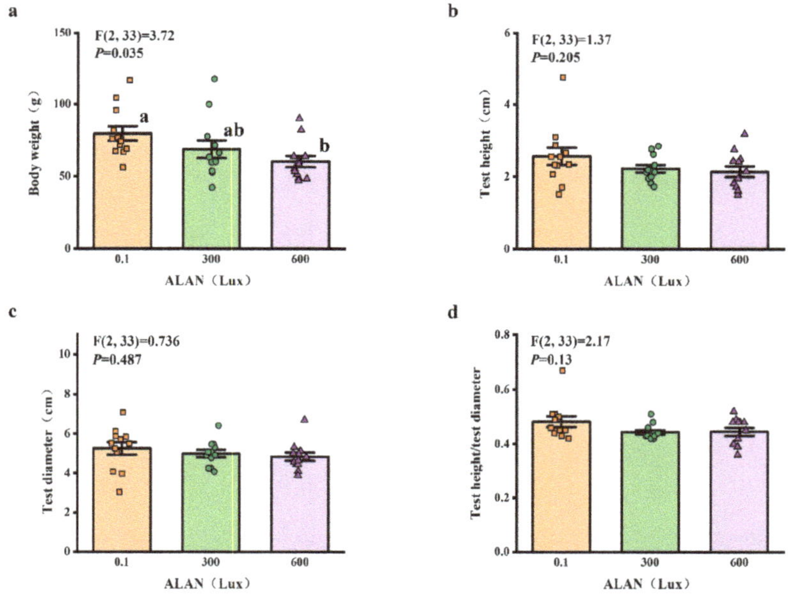

Figure 4. Effects of ALAN at different intensities for 6 weeks on body weight (**a**), test height (**b**), test diameter (**c**), and test height/test diameter (**d**) of sea urchins. Different letters indicate significant differences. Values are shown in data points and mean ± S.E. ($n = 12$).

3.5. Lantern Weight and Length

We found that Aristotle's lantern weight of *H. crassispina* significantly decreased exposure to ALAN at 300 Lux and 600 Lux compared to those exposed to ALAN at 0.1 ($p < 0.05$; $p < 0.05$). However, there was no significant difference between the individuals exposed to 0.1 and 300 Lux ($p = 0.906$; Figure 5a). In addition, Aristotle's lantern length among the individuals exposed to 0.1, 300, and 600 Lux of ALAN showed no significant difference after 6 weeks of the experiment (Figure 5b).

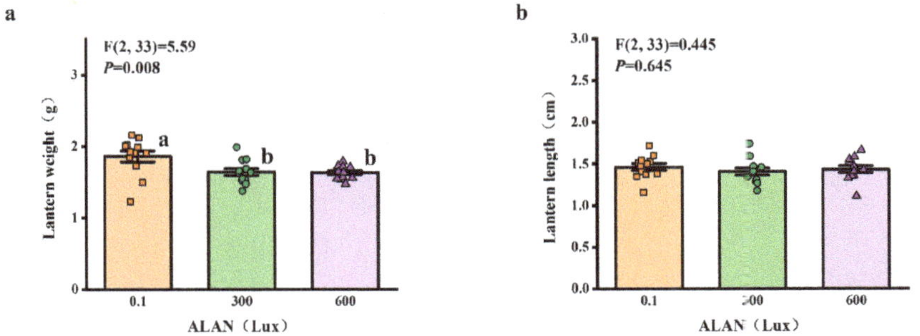

Figure 5. Effects of ALAN at different intensities for 6 weeks on Aristotle's lantern weight (**a**) and Aristotle's lantern length (**b**) of sea urchins. Different letters indicate significant differences. Values are shown in data points and mean ± S.E. ($n = 12$).

3.6. Gonad and Gut-Weight

We found that there was a significant difference in GSI (gonads weight/test weight) among the individuals exposed to 0.1, 300, and 600 Lux of ALAN (all $p < 0.05$; Figure 6a). There was no significant difference in gut-weight among the individuals exposed to 0.1, 300, and 600 Lux of ALAN (Figure 6b).

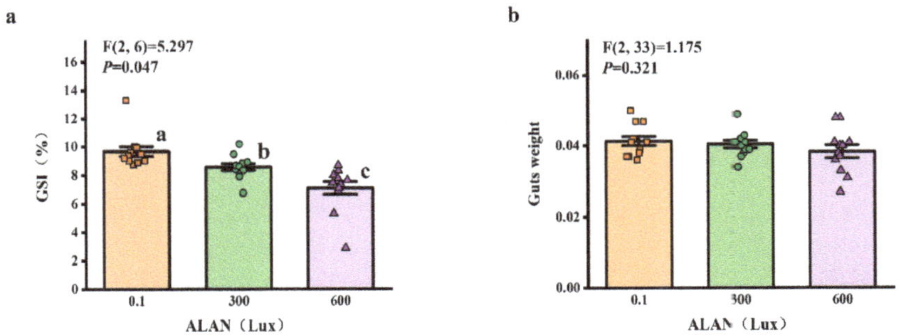

Figure 6. Effects of ALAN at different intensities for 6 weeks on GSI (**a**) and gut-weight (**b**) of sea urchins. Different letters indicate significant differences. Values are shown in data points and mean ± S.E. ($n = 12$).

3.7. 5-HIAA/5-HT Ratio

We used the 5-HIAA/5-HT ratio as an indicator of serotonin system activity, and we found that ALAN at 600 Lux significantly increased the 5-HIAA/5-HT ratio of *H. crassispina* compared to those exposed to ALAN at 0.1 and 300 Lux of ALAN ($p < 0.005$; $p < 0.005$). However, there was no significant difference between the individuals exposed to 0.1 and 300 Lux of ALAN ($p = 0.054$; Figure 7).

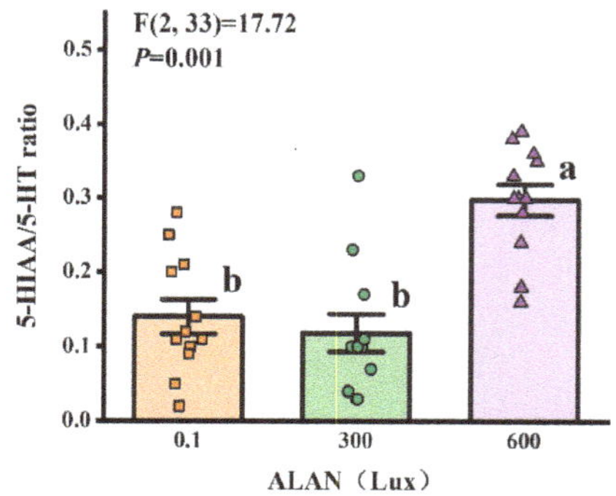

Figure 7. Effects of ALAN at different intensities for 6 weeks on coelomic fluid 5-HIAA/5-HT ratio of sea urchins. Different letters indicate significant differences. Values are shown in data points and mean ± S.E. (n = 12).

3.8. *Pax6* Gene Expression

We found that *Pax6* gene expression was significantly affected by ALAN at 600 Lux in *H. crassispina* compared with the control (0.1 Lux) ($p < 0.005$; $p < 0.005$). No significant difference in *Pax6* gene expression was found between the individuals exposed to 0.1 and 300 Lux of ALAN (p = 0.192; Figure 8).

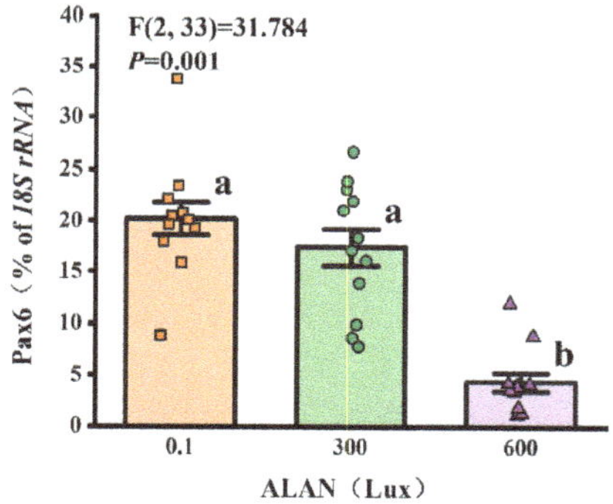

Figure 8. Effects of ALAN at different intensities for 6 weeks on *Pax6* gene expression of sea urchins. Different letters indicate significant differences. Values are shown in data points and mean ± S.E. (n = 12).

4. Discussion

In this work, we constructed small-scale systems to investigate the effects of artificial light at night (ALAN) on fitness-related traits of *H. crassispina*. The main results were as follows: (1) ALAN at 600 Lux significantly increased the righting response time, the covering response time, and the 5-HIAA/5-HT ratio, and decreased food consumption, body weight, Lantern weight, GSI, and *Pax6* gene expression; (2) ALAN at 300 Lux significantly decreased food consumption, Lantern weight, and GSI.

4.1. *Effects of ALAN on Behavioral Responses*

Complex marine benthic environments shape a number of ecologically important behaviors in sea urchins. Covering and righting are two representative behaviors with fitness importance in antipredation [21–25]. The diminished covering and righting capacities of *H. crassispina* observed in this study suggest that extended exposure to ALAN might unfavorably impact fitness-related behaviors. This behavioral shift was linked to the decrease in sea urchin populations, potentially attributed to heightened predation pressure within the coastal ecosystem [28]. Similar outcomes highlighting the influence of

ALAN on fitness-related behaviors have been documented in diverse coastal organisms. For instance, sea turtles (*Caretta caretta* and *Chelonia mydas*) may fail to navigate toward the ocean post-hatching due to ALAN-induced disruptions in their behavioral orientation [44]. Seabird migratory patterns also exhibit alteration, as birds are drawn to and linger around artificial light sources [45–47]. The composition of shallow water fish species underwent modification under ALAN influence: large predatory and small shoaling fish congregated around illuminated zones during the night, leading to reduced prey availability [30]. Notably, a study involving mud snails (*Hydrobia ulvae*) on French mudflats unveiled a decline in the crawling activity (and growth) of snails in response to light exposure [48].

The formation of metabolites like 5-HIAA arises subsequent to the re-uptake of 5-HT from the synaptic cleft, and their accumulation in neural tissue likely exhibits a time-dependent pattern [36]. Consequently, evaluating tissue concentrations of neurotransmitter metabolites fails to provide an immediate depiction of neural activity. As such, the ratio of metabolite to monoamine was adopted as a marker indicating heightened monoamine utilization. In this study, the 5-HIAA/5-HT ratio in *H. crassispina* exhibited an increase following prolonged ALAN exposure. This observation suggests that the serotonin system was stimulated in response to light pollution, causing a progressive augmentation in metabolite levels over time. Numerous reports have delved into serotonin's influence on animal behavior, consistently revealing its inhibitory effects on traits such as aggressive behavior, feeding patterns, and motor activity [36,49]. Therefore, alterations in the concentration of the monoamine transmitter serotonin may underlie the internal mechanism driving the behavioral shifts observed in sea urchins within this study.

4.2. Effects of ALAN on Growth

H. crassispina subjected to prolonged artificial light exposure displayed a pronounced reduction in food consumption compared to those shielded from ALAN. This correlation is substantiated by our finding that ALAN significantly suppressed Aristotle's lantern reflex, a pivotal factor in sea urchin food intake. This diminished food consumption provides a plausible explanation for the subsequent reduction in body size. These findings strongly indicate that light pollution could exert an impact on the growth of intertidal organisms. Correspondingly, studies by *Luarte* et al. involving both laboratory and field investigations demonstrated that the foraging behavior and growth rate of the amphipod *Orchestoidea tuberculate* were significantly affected by ALAN [50]. Field observations indicated that lower light levels (60 Lux) curtailed amphipod feeding and growth rates. The growth rates of amphipods under ambient conditions (darkness) were nearly three times higher compared to those exposed to ecological light pollution. Similarly, under ambient conditions, the food consumption rates were nearly two times higher than under exposure to ecological light pollution. The potential influence of light pollution on growth and development also extends to mollusks. Research unveiled a correlation between daily light exposure duration and the growth pace and stored energy levels in the pulmonate *Lymnaea stagnalis* [51]. These outcomes find alignment with investigations on rodents, specifically *Myotis emarginatus*, showcasing that light pollution diminishes consumption rates and impedes the growth of juvenile and suckling bats [52]. Thus, the decreased food consumption stemming from light pollution is likely to curtail the fitness of littoral organisms by reducing energy reserves.

4.3. Effects of ALAN on Reproductive Performance

Among the pivotal factors influencing sea urchin reproduction, food consumption holds paramount significance [20,21]. To gauge reproductive performance, we employed the Gonad-Somatic Index (GSI) and identified a GSI reduction in *H. crassispina* exposed to ALAN, which correlated with diminished food intake. This could suggest that sea urchins possibly allocate nutrients and energy to combat environmental stress, consequently compromising gonad tissue development. Furthermore, ALAN exhibited no significant influence on gut-weight in *H. crassispina*, a finding consistent with previous research [21]. This implies that sea urchin gonads are potentially more susceptible to environmental stressors than their gastrointestinal counterparts. Thus, ALAN exposure might detrimentally impact the reproductive capacity of sea urchins.

The development of gonad tissue follows a hormonal cascade, with gonadotropin-releasing hormone (GnRH) triggering the release of luteinizing hormone (LH) and follicle-stimulating hormone (FSH) from the pituitary gland, leading to the production of sex steroids in vertebrates [53]. The regulatory mechanism governing gonadal development is less defined in invertebrate sea urchins. As a result, we did not investigate relevant hormones to probe the mechanism through which light pollution affects sea urchin reproduction. Nonetheless, analogous studies have been conducted in marine fish. For instance, in female *Lateolabrax maculates*, mRNA expression of luteinizing hormone and follicle-stimulating hormone was suppressed even at white light levels as low as 1 Lux at night [54]. Similarly, a study in a natural setting found diminished mRNA expression of gonadotropins (luteinizing hormone and follicle-stimulating hormone) and circulating sex hormones (17β-oestradiol and 11-ketotestosterone) in European perch under street lighting conditions (13.3–16.5 Lux at the surface) [55]. Consequently, exploring the internal mechanism of light's impact on gonad development in invertebrates such as sea urchins should be a prospective avenue of research.

4.4. Effects of ALAN on Pax6 Gene Expression

In a pioneering effort, we successfully isolated and sequenced a complete cDNA encoding a 440-amino-acid *Pax6* from the tubular feet of *H. crassispina*. We conducted an in-depth analysis of *Pax6*'s similarity and evolutionary relationships across various species (Supplementary Materials: Table S2, Figure S1–S3), thereby furnishing fundamental biological insights for subsequent investigations into the *Pax6* gene within sea urchins. Notably, *Pax6* stands as a transcription factor gene with established involvement in eye development and photoreception in ocular-forming organisms [56]. The presence of *Pax6* expression signifies the preservation of photoreceptor cell specification processes, functioning as a canonical eye gene even in creatures devoid of specialized eyes [57]. Consequently, the *Pax6* gene emerges as a central focus for probing the photosensitive systems responding to the light milieu in sea urchins. For instance, the relative *Pax6* expression in *Strongylocentrotus intermedius* displayed a negative correlation with natural light intensity. However, *Pax6* expression patterns in *S. intermedius* underwent alterations following UV-B radiation exposure [58]. In line with this, our present experiment reveals that prolonged ALAN exposure also disrupts the typical expression of the *pax6* gene in *H. crassispina*. This suggests that persistent light stimulation might perturb molecules inherent to the photosensitive systems of sea urchins, ultimately yielding anomalous gene expression. On the whole, *Pax6* gene expression is poised to function as a sensitive gauge for assessing potential damage to the photosensitive system of sea urchins.

4.5. Sea Urchins in the Wild Light Environment

The profound impact of nocturnal light on the fitness traits of sea urchins can be attributed to their inherent nocturnal activity patterns [2]. Continuous exposure to light inhibits sea urchin activity, consequently slowing down behaviors such as covering and righting, as well as impairing their feeding efficiency. Additionally, this light exposure also leads to changes in hormone levels and gene expression within their bodies. This strong correlation between nocturnal light and the behavioral responses of sea urchins highlights the significant role that light plays in regulating their activity levels and overall ecological dynamics [31]. Despite this, considering the complexity of intertidal habitats, the impact of nocturnal light on sea urchins in the wild may not be consistent, especially in rocky coastal areas where cracks, rocks, and shade-providing seaweeds are present. Sea urchins may opt to inhabit shaded areas or regions with suitable light intensity. For instance, a study conducted on a drumfish species found that their activity range shifted in response to nighttime light exposure of 70 Lux [59]. However, the narrowing of the activity range was determined by the encroachment of intensifying nocturnal light into dark spaces. Therefore, it is crucial to adequately understand the complexity of natural settings. Further research conducted in the wild, including the use of large mesocosms, or in more natural laboratory settings, will be necessary.

5. Conclusions

These findings illuminate the potential detrimental impact of extended (6-week) exposure to ALAN on the behavioral responses, growth, reproductive performance, and photoreception of sea urchins. Importantly, the study findings reveal that the sea urchins' responses to light are more pronounced when exposed to higher light intensities, emphasizing the increased threat posed by intensified artificial light at night (ALAN) to their overall fitness. This highlights the potential risks that intertidal coastal ecosystems face in areas characterized by prominent luminous installations, such as stadiums, ports, and wharves. These three locations, due to their excessive lighting, emerge as significant sources of disturbance and risk for the delicate balance of these ecosystems. Consequently, it is crucial to consider the conservation and management of these areas to mitigate the adverse effects of ALAN and preserve the ecological health and biodiversity of intertidal coastal environments. This study significantly advances our comprehension of the fitness-related attributes of sea urchins under ALAN exposure, furnishing crucial insights into the ecological risks of coastal regions.

Supplementary Materials: The following supporting information can be downloaded at: https://www.mdpi.com/article/10.3390/ani13193035/s1, Figure S1: The deduced amino acid sequences of *Pax6* from *H. crassispina*, Figure S2: Alignment of deduced amino acid sequences of *Pax6* from *H. crassispina* and select species, Figure S3: Phylogenetic tree illustrating the relationship of *Pax6* from *H. crassispina* and select species, Table S1: Water quality conditions during the experiment; *Procedure (Sp)*: Cloning and analysis of full-length *Pax6* gene, Table S2: Nucleotide sequences of the primers for PCR.

Author Contributions: Conceptualization, T.W.; investigation, X.X. and Z.W.; data curation, X.X. and K.D.; validation, J.Y.; format analysis, X.X. and Z.W.; resources, X.X. and X.J.; writing—original draft preparation, X.X. and T.W.; writing—review and editing, X.X., T.W. and J.Y.; project administration, T.W; funding acquisition, X.X. and T.W. All authors have read and agreed to the published version of the manuscript.

Funding: This research was funded by the "Pioneer" and "Leading Goose" R&D Program of Zhejiang (2022C02040) and the Science and Technology Program of Zhoushan (2022C41016).

Institutional Review Board Statement: Ethical review and approval were waived for this study due to no authorization being required for experimental studies on invertebrates such as sea urchins. The experiment with local laws and regulations, adheres to basic ethical principles, and ensures respect and responsible utilization of the animals.

Informed Consent Statement: Not applicable.

Data Availability Statement: The data that support the findings of this study are available upon reasonable request from the corresponding author.

Acknowledgments: The authors also thank the Fisheries Research Institute of Zhoushan Zhejiang for their support and help in providing data and sample collection throughout the study.

Conflicts of Interest: The authors declare no conflict of interest.

References

1. Gaston, K.J.; Bennie, J.; Davies, T.W.; Hopkins, J. The ecological impacts of nighttime light pollution: A mechanistic appraisal. *Biol. Rev.* **2013**, *88*, 912–927. [CrossRef] [PubMed]
2. Zapata, M.J.; Sullivan, S.M.P.; Sullivan, S.M. Artificial lighting at night in estuaries implications from individuals to ecosystems. *Estuar. Coast.* **2019**, *42*, 309–330. [CrossRef]
3. Bassi, A.; Love, O.P.; Cooke, S.J.; Warriner, T.R.; Harris, C.M.; Madliger, C.L. Effects of artificial light at night on fishes: A synthesis with future research priorities. *Fish Fish.* **2022**, *23*, 631–647. [CrossRef]

4. Davies, T.W.; Duffy, J.P.; Bennie, J.; Gaston, K.J. Stemming the tide of light pollution encroaching into marine protected areas. *Conserv. Lett.* **2016**, *9*, 164–171. [CrossRef]
5. Davies, T.W.; Smyth, T. Why artificial light at night should be a focus for global change research in the 21st century. *Global Change Biol.* **2018**, *24*, 872–882. [CrossRef]
6. Marangoni, L.F.B.; Davies, T.; Smyth, T.; Rodríguez, A.; Hamann, M.; Duarte, C.; Pendoley, K.; Berge, J.; Maggi, E.; Levy, O. Impacts of artificial light at night in marine ecosystems-A review. *Global Change Biol.* **2022**, *28*, 5346–5367. [CrossRef]
7. Hussein, A.A.A.; Bloem, E.; Fodor, I.; El-Sayed, B.; Tadros, M.M.; Soliman, M.F.M.; El-Shenawy, N.S.; Koene, J.M. Slowly seeing the light: An integrative review on ecological light pollution as a potential threat for mollusks. *Environ. Sci. Pollut. Res. Int.* **2021**, *28*, 5036–5048. [CrossRef]
8. Davies, T.W.; Duffy, J.P.; Bennie, J.; Gaston, K.J. The nature, extent, and ecological implications of marine light pollution. *Front. Ecol. Environ.* **2014**, *12*, 347–355. [CrossRef]
9. Russart, K.L.G.; Nelson, R.J. Artificial light at night alters behavior in laboratory and wild animals. *J. Exp. Zool. Part A* **2018**, *329*, 401–408. [CrossRef]
10. Ayalon, I.; Rosenberg, Y.; Benichou, J.I.C.; Campos, C.L.D.; Sayco, S.L.G.; Nada, M.A.L.; Baquiran, J.I.P.; Ligson, C.A.; Avisar, D.; Conaco, C.; et al. Coral gametogenesis collapse under artificial light pollution. *Curr. Biol.* **2021**, *31*, 413–419. [CrossRef]
11. Colman, L.P.; Lara, P.H.; Bennie, J.; Broderick, A.C.; de Freitas, J.R.; Marcondes, A.; Witt, M.J.; Godley, B.J. Assessing coastal artificial light and potential exposure of wildlife at a national scale: The case of marine turtles in Brazil. *Biodivers. Conserv.* **2020**, *29*, 1135–1152. [CrossRef]
12. Shimada, T.; Meekan, M.G.; Baldwin, R.; Al-Suwailem, A.M.; Clarke, C.; Santillan, A.S.; Duarte, C.M. Distribution and temporal trends in the abundance of nesting sea turtles in the Red Sea. *Biol. Conserv.* **2021**, *261*, 109235. [CrossRef]
13. Ronconi, R.A.; Allard, K.A.; Taylor, P.D. Bird interactions with offshore oil and gas platforms: A review of impacts and monitoring techniques. *J. Environ. Manag.* **2015**, *147*, 34–45. [CrossRef]
14. Maboloc, E.A.; Fang, J.K.H. Tissue regeneration of the purple sea urchin *Heliocidaris crassispina*. *Bull. Mar. Sci.* **2023**, *99*, 19–20. [CrossRef]
15. Kanamoto, Y.; Nakamura, K. Age and growth analyses of the purple sea urchin *Heliocidaris crassispina* inhabiting different feeding environments in the Shimane Peninsula, Japan. *Reg. Stud. Mar. Sci.* **2023**, *65*, 103096. [CrossRef]
16. Kadota, T.; Kiyomoto, S.; Masuda, Y.; Miyano, T.; Yoshimura, T. Restoration of a small-sized macroalgal bed through the removal of sea urchins in Kashiyama, Nagasaki Prefecture. *Nippon. Suisan Gakk.* **2022**, *88*, 49–57. [CrossRef]
17. Feng, W.P.; Nakabayashi, N.; Inomata, E.; Inomata, E.; Aoki, M.N.; Agatsuma, Y. Sexually unbalanced gonad development and nutrition of the newly range-extended sea urchin *Heliocidaris crassispina* in the northeastern Honshu, Japan. *Estuar. Coast. Shelf S.* **2021**, *249*, 107120. [CrossRef]
18. Ishii, M.; Unuma, T.; Masadate, A.; Hoshikawa, H.; Takahashi, K.; Kosaka, S.; Masuda, A.; Murakami, K. Accelerated photoperiod promotes gonadal maturation in the sea urchin *Strongylocentrotus intermedius*. *Fish. Sci.* **2022**, *88*, 299–310. [CrossRef]
19. Feng, W.P.; Nakabayashi, N.; Inomata, E.; Aoki, M.N.; Agatsuma, Y. Impacts of water temperature on the physiology and behaviours of the sea urchins *Heliocidaris crassispina* and *Mesocentrotus nudus* that reflect their range extension and disappearance in the Oga Peninsula, northern Honshu, Japan. *Can. J. Fish. Aquat. Sci.* **2020**, *78*, 580–588. [CrossRef]
20. Wang, X.B.; Li, X.S.; Xiong, D.Q.; Chen, H.S.; Ren, H. Effects of stranded heavy fuel oil subacute exposure on the fitness-related traits of sea urchin *Strongylocentrotus intermedius*. *Mar. Freshwater Res.* **2022**, *73*, 754–761. [CrossRef]
21. Zhao, C.; Zhang, L.L.; Shi, D.T.; Chi, X.M.; Yin, D.H.; Sun, J.N.; Ding, J.Y.; Yang, M.F.; Chang, Y.Q. Carryover effects of short-term UV-B radiation on fitness related traits of the sea urchin *Strongylocentrotus intermedius*. *Ecotox. Environ. Safe.* **2018**, *164*, 659–664. [CrossRef] [PubMed]
22. Wei, J.; Zhang, L.S.; Zhao, C.; Feng, W.P.; Sun, P.; Chang, Y.Q. Correlation analyses of covering and righting behaviors to fitness related traits of the sea urchin *Glyptocidaris crenularis* in different environmental conditions. *Chin. J. Oceanol. Limn.* **2016**, *34*, 1183–1190. [CrossRef]
23. Verling, E.; Crook, A.; Barnes, D. Covering behaviour in *Paracentrotus lividus*: Is light important? *Mar. Biol.* **2002**, *140*, 391–396. [CrossRef]
24. Chi, X.; Shi, D.; Ma, Z.; Hu, F.Y.; Sun, J.N.; Huang, X.Y.; Zhang, L.S.; Chang, Y.Q.; Zhao, C. Carryover effects of long-term high water temperatures on fitness-related traits of the offspring of the sea urchin *Strongylocentrotus intermedius*. *Mar. Environ. Res.* **2021**, *169*, 105371. [CrossRef] [PubMed]
25. Pawson, D.L.; Pawson, D.J. Bathyal sea urchins of the Bahamas, with notes on covering behavior in deep sea echinoids (Echinodermata: Echinoidea). *Deep Sea Res. Part II* **2013**, *92*, 207–213. [CrossRef]
26. Backström, T.; Winberg, S. Serotonin coordinates responses to social stress-what we can learn from fish. *Front. Neurosci.* **2017**, *11*, 595. [CrossRef]
27. Zhao, C.; Ji, N.J.; Tian, X.F.; Feng, W.P.; Sun, P.; Wei, J. *Opsin4*, *Opsin5*, and *Pax6* significantly increase their expression in recently settled juveniles of the sea urchin *Strongylocentrotus intermedius* (Echinodermata: Echinoidea). *Inver. Reprod. Dev.* **2015**, *5*, 119–123. [CrossRef]
28. Belleza, D.F.C.; Kawabata, Y.; Toda, T.; Gregory, N.N. Effects of dead conspecifics, hunger states, and seasons on the foraging behavior of the purple urchin *Heliocidaris crassispina*. *Mar. Ecol. Prog. Ser.* **2020**, *664*, 133–148. [CrossRef]

29. Vowles, A.S.; Kemp, P.S. Artificial light at night (ALAN) affects the downstream movement behaviour of the critically endangered European eel, *Anguilla anguilla*. *Environ. Pollut.* **2021**, *274*, 116585. [CrossRef]
30. Pulgar, J.; Manríquez, P.H.; Widdicombe, S.; García-Huidobro, R.; Quijón, P.A.; Carter, M.; Aldana, M.; Quintanilla-Ahumada, D.; Duarte, C. Artificial Light at Night (ALAN) causes size-dependent effects on intertidal fish decision-making. *Mar. Pollut. Bull.* **2023**, *193*, 115190. [CrossRef]
31. Fobert, E.K.; Burke da Silva, K.; Swearer, S.E. Artificial light at night causes reproductive failure in clownfish. *Biol. Lett.* **2019**, *15*, 20190272. [CrossRef] [PubMed]
32. Kupprat, F.; Kloas, W.; Krüger, A.; Schmalsch, C.; Hölker, F. Misbalance of thyroid hormones after two weeks of exposure to artificial light at night in Eurasian perch *Perca fluviatilis*. *Conserv. Physiol.* **2021**, *9*, coaa124. [CrossRef] [PubMed]
33. Newman, R.C.; Ellis, T.; Davison, P.I.; Ives, M.J.; Thomas, R.J.; Griffiths, S.W.; Riley, W.D. Using novel methodologies to examine the impact of artificial light at night on the cortisol stress response in dispersing Atlantic salmon (*Salmo salar* L.) fry. *Conserv. Physiol.* **2015**, *3*, cov051. [CrossRef]
34. Wang, H.; Ding, J.; Ding, S.; Chang, Y. Transcriptome analysis to characterize the genes related to gonad growth and fatty acid metabolism in the sea urchin *Strongylocentrotus intermedius*. *Genes Genom.* **2019**, *41*, 1397–1415. [CrossRef] [PubMed]
35. Terwilliger, D.P.; Buckley, K.M.; Brockton, V.; Ritter, N.J.; Smith, L.C. Distinctive expression patterns of 185/333 genes in the purple sea urchin, *Strongylocentrotus purpuratus*: An unexpectedly diverse family of transcripts in response to LPS, beta-1,3-glucan, and dsRNA. *BMC Mol. Biol.* **2007**, *8*, 16. [CrossRef] [PubMed]
36. Xu, X.; Zhang, Z.; Guo, H.; Qin, J.; Zhang, X. Changes in aggressive behavior, cortisol and brain monoamines during the formation of social hierarchy in black rockfish (*Sebastes schlegelii*). *Animals* **2020**, *10*, 2357. [CrossRef] [PubMed]
37. Xu, X.; Sun, X.; Bai, Q.; Zhang, Y.; Qin, J.; Zhang, X. Molecular identification of an androgen receptor and the influence of long-term aggressive interaction on hypothalamic genes expression in black rockfish (*Sebastes schlegelii*). *J. Comp. Physiol. A* **2021**, *207*, 401–413. [CrossRef] [PubMed]
38. Xu, X.; Guo, H.; Zhang, Z.; Wang, Y.; Qin, J.; Zhang, X. Impact of pre-aggressive experience on behavior and physiology of black rockfish (*Sebastes schlegelii*). *Aquaculture* **2021**, *536*, 736416. [CrossRef]
39. Zhang, Z.; Bai, Q.; Xu, X.; Zhang, X. Effects of the dominance hierarchy on social interactions, cortisol level, HPG-axis activities and reproductive success in the golden cuttlefish *Sepia esculent*. *Aquaculture* **2020**, *533*, 736059. [CrossRef]
40. Yang, M.; Yu, Y.; Ding, J.; Sun, J.; Hu, F.; Chang, Y.; Zhao, C. Effects of light intensity on *Opsin4*, *Opsin5*, and *Pax6* expressions of the sea urchin *Strongylocentrotus intermedius*. *Mar. Ecol.* **2020**, *41*, e12593. [CrossRef]
41. Parra, M.; Rubilar, T.; Latorre, M.P.; Parra, M.; Rubilar, T.; Latorre, M.P.; Ephedra, L.; Gil, D.G.; Díaz de Vivar, M.E. Nutrient allocation in the gonads of the sea urchin *Arbacia dufresnii* in different stages of gonadal development. *Invertebr. Reprod. Dev.* **2015**, *59*, 26–36. [CrossRef]
42. Machado, A.M.; Fernández-Boo, S.; Nande, M.; Pinto, R.; Costas, B.; Castro, L.F. The male and female gonad transcriptome of the edible sea urchin, *Paracentrotus lividus*: Identification of sex-related and lipid biosynthesis genes. *Aquacult. Rep.* **2022**, *22*, 100936. [CrossRef]
43. Zhang, B.; Yang, J.W.; Han, T.; Huang, D.X.; Zhao, Z.H.; Feng, J.Q.; Zhou, N.M.; Xie, H.Q.; Wang, T.M. Identification and characterization of a novel 5-hydroxytryptamine receptor in the sea cucumber *Apostichopus japonicus* (Selenka). *J. Exp. Zool. Part A* **2021**, *335*, 367–380. [CrossRef] [PubMed]
44. Tuxbury, S.M.; Salmon, M. Competitive interactions between artificial lighting and natural cues during seafinding by hatchling marine turtles. *Biol. Conserv.* **2005**, *121*, 311–316. [CrossRef]
45. La Sorte, F.A.; Fink, D.; Butler, J.J.; Farnsworth, A.; Cabrera-Cruz, S.A. Seasonal associations with urban light pollution for nocturnally migrating bird populations. *Glob. Change Biol.* **2017**, *23*, 4609–4619. [CrossRef]
46. Van Doren, B.M.; Horton, K.G.; Dokter, A.M.; Klinck, H.; Elbin, S.B.; Farnsworth, A. High-intensity urban light installation dramatically alters nocturnal bird migration. *Proc. Natl. Acad. Sci. USA* **2017**, *114*, 11175–11180. [CrossRef]
47. Burt, C.S.; Kelly, J.F.; Fox, A.S.; Jenkins-Smith, H.C.; Leon-Corwin, M.; Khalighifar, A.; Trankina, G.E.; Silva, C.L.; Horton, K.G. Can ecological forecasting lead to convergence on sustainable lighting policies? *Conserv. Sci. Pract.* **2023**, *5*, e12920. [CrossRef]
48. Orvain, O.; Sauriau, P.G. Environmental and behavioural factors affecting activity in the intertidal gastropod *Hydrobia ulvae*. *J. Exp. Mar. Biol. Ecol.* **2002**, *272*, 191–216. [CrossRef]
49. Morandini, L.; Ramallo, M.R.; Scaia, M.F.; Höcht, C.; Somoza, G.M.; Pandolfi, M. Dietary L-tryptophan modulates agonistic behavior and brain serotonin in male dyadic contests of a cichlid fish. *J. Comp. Physiol. A* **2019**, *205*, 867–880. [CrossRef]
50. Luarte, T.; Bonta, C.C.; Silva-Rodriguez, E.A.; Quijón, P.A.; Miranda, C.; Fariasm, A.A.; Duarte, C. Light pollution reduces activity, food consumption and growth rates in a sandy beach invertebrate. *Environ. Pollut.* **2016**, *218*, 1147–1153. [CrossRef]
51. Hussein, A.A.A.; Baz, E.S.; Mariën, J.; Tadros, M.M.; El-Shenawy, N.S.; Koene, J.M. Effect of photoperiod and light intensity on learning ability and memory formation of the pond snail *Lymnaea stagnalis*. *Invert. Neurosci.* **2020**, *20*, 18. [CrossRef]
52. Boldogh, S.; Dobrosi, D.; Samu, P. The effects of the illumination of buildings on house-dwelling bats and its conservation consequences. *Acta. Chiropterol.* **2007**, *9*, 527–534. [CrossRef]
53. Silvia, L.; Manuel, T.S. Dissecting the roles of gonadotropin-inhibitory hormone in mammals: Studies using pharmacological tools and genetically modified mouse models. *Front. Endocrinol.* **2015**, *6*, 189. [CrossRef]
54. Brüning, A.; Hölker, F.; Franke, S.; Kleiner, W.; Kloas, W. Impact of different colours of artificial light at night on melatonin rhythm and gene expression of gonadotropins in European perch. *Sci. Total. Environ.* **2016**, *543*, 214–222. [CrossRef]

55. Brüning, A.; Kloas, W.; Preuer, T.; Hölker, F. Influence of artificially induced light pollution on the hormone system of two common fish species, perch and roach, in a rural habitat. *Conserv. Physiol.* **2018**, *6*, coy016. [CrossRef]
56. Raible, F.; Tessmar-Raible, K.; Arboleda, E.; Kaller, T.; Bork, P.; Arendt, D.; Arnone, M.I. Opsins and clusters of sensory G-protein-coupled receptors in the sea urchin genome. *Dev. Biol.* **2006**, *300*, 461–475. [CrossRef] [PubMed]
57. Ullrich-Luter, E.M.; Dupont, S.; Arboleda, E.; Hausen, H.; Arnone, M.I. Unique system of photoreceptors in sea urchin tube feet. *Proc. Natl. Acad. Sci. USA* **2011**, *108*, 8367–8372. [CrossRef] [PubMed]
58. Zhao, C.; Ji, N.; Sun, P.; Feng, W.; Wei, J.; Chang, Y. Effects of light and covering behavior on *PAX6* expression in the sea urchin *Strongylocentrotus intermedius*. *PLoS ONE* **2014**, *9*, e110895. [CrossRef] [PubMed]
59. Pulgar, J.; Zeballos, D.; Vargas, J.; Aldana, M.; Manriquez, P.H.; Manriquez, K.; Quijón, P.A.; Widdicombe, S.; Anguita, C.; Quintanilla, D.; et al. Endogenous cycles, activity patterns and energy expenditure of an intertidal fish is modified by artificial light pollution at night (ALAN). *Environ. Pollut.* **2019**, *244*, 361–366. [CrossRef]

animals

MDPI

Review

Ethical Considerations for Echinoderms: New Initiatives in Welfare

Augusto César Crespi-Abril [1,2,*] and Tamara Rubilar [2,3]

1 Instituto Patagónico del Mar (IPaM), Universidad Nacional del a Patagonia San Juan Bosco (UNPSJB), Boulevard Brown 2915, Puerto Madryn 9120, Argentina

2 Laboratorio de Oceanografía Biológica (LOBio), Centro Para el Estudio de Sistemas Marinos (CESIMAR–CONICET), Boulevard Brown 2915, Puerto Madryn 9120, Argentina; rubilar@cenpat-conicet.gob.ar or crubilar@unpata.edu.ar

3 Laboratorio de Química de Organismos Marinos (LabQuiOM), Instituto Patagónico del Mar (IPAM), Facultad de Ciencias Naturales y Ciencias de la Salud, Universidad Nacional de la Patagonia San Juan Bosco, Boulevard Brown 2930, Puerto Madryn 9120, Argentina

* Correspondence: crespi@cenpat-conicet.gob.ar or acrespi@unpata.edu.ar

Simple Summary: This study discusses the moral issues surrounding the study of echinoderms, a class of marine animals that is attracting interest from scientists. The significance of safely handling these animals is stressed and promotes an ethical standard that is customized to their needs. Our strategy was heavily influenced by the 3Rs principle, which advocates for substituting aware living vertebrates with non-sentient material in research. Echinoderms are excellent models for experimental inquiry because they are typically thought to be non-sentient. Although it is not possible to assess their mental states at this time, there is ample evidence of social behavior in many species, suggesting that ignoring interactions with them could be detrimental to their wellbeing. Recently, progress has been made toward developing an ethical framework for invertebrates, such as crustaceans, echinoderms, and cephalopods. To protect the welfare of echinoderms even in the absence of specific standards, we suggest an enlarged version of the 5Rs framework that includes responsibility and respect. In addition to advancing our knowledge of these interesting species, this research establishes a critical standard for the responsible and sympathetic care of all animals in scientific research.

Abstract: This paper explores the ethical considerations surrounding research on echinoderms, a group of invertebrates that has recently garnered attention in the scientific community. The importance of responsible animal handling and the need for an ethical framework that encompasses echinoderms are emphasized. The 3Rs principle, advocating for the replacement of conscious living vertebrates with non-sentient material in research, is discussed as a guiding tool in current animal research practices. As invertebrates are generally classified as non-sentient animals, the replacement dimension tends to favor them as prevalent models in experimental research. While it currently lacks the means to assess the mental states of invertebrates, there is undeniable evidence of social behavior in many species, suggesting that a lack of interactions with these organisms could potentially adversely affect their wellbeing. In the last few years, considerable progress has been made in developing an ethical framework that takes invertebrates into account, particularly cephalopods, crustaceans, and echinoderms. In this context, we discuss the development of a broader conceptual framework of 5Rs that includes responsibility and respect, which may guide practices ensuring welfare in echinoderms, even in the absence of any particular normative.

Keywords: echinoderm welfare; 5Rs principle; respect; responsibility; invertebrates; responsible echinoderm use; echinoderm sentience

1. Introduction

Animal use and interaction have played a crucial role in human endeavors throughout recorded human history. Food, transportation, research (primarily in the field of medical investigation), clothing, and companionship are among the main applications. Notably, this engagement includes invertebrates, indicating their natural involvement in various aspects of human existence. While some aspects of these relationships are unequivocally positive, such as providing sustenance, serving as research models, or providing companionship, others are negative, lacking in purpose, or even inflict harm. This dichotomy may stem from the classification of certain invertebrates as pests or vectors of human diseases. Such adverse interactions have elicited widespread aversion or apprehension toward numerous invertebrate species [1]. As a result, ethical concerns about these creatures must be addressed to ensure their viability as experimental modeling alternatives to vertebrates.

Invertebrates account for more than 90% of the total biodiversity on Earth [1]. This vast biological realm includes 36 phyla of invertebrates, 8 of which share the most common associations with humans: Porifera, Cnidaria, Platyhelminthes, Nematoda, Annelida, Arthropoda (the largest phylum in the animal kingdom), Mollusca (the second-largest phylum in the animal kingdom), and Echinodermata [1]. Although these phyla are all classified as invertebrates, their diversity is nothing short of astounding. Morphological, nervous system, and behavioral characteristics are unique to each phylum and can even vary within one [2–4].

Recently, there has been a shift in the ethical consideration of invertebrates, prompting a number of studies to lay the philosophical groundwork for incorporating invertebrates into ethical discourse [5–8]. The intricate behaviors of specific invertebrates, particularly octopuses (Cephalopods), have primarily prompted this transformation. Close interaction with octopuses in aquarium settings has aided in the identification of individuality and observable behaviors (personality traits) in these creatures [7,9]. Despite their distinct nervous system configuration from that of vertebrates, these studies have revealed octopuses' remarkable intelligence and sentience [10]. Furthermore, in 2013, cephalopods were included in European Union legislation regarding the protection of animals used for scientific purposes, putting them on par with vertebrates (EU, 2010, Directive 2010/63/EU). This pivotal advancement not only adds to the ethical consideration for invertebrates in general, but it also represents a pivotal advancement in the field [3,8,11]. It is paramount to recognize that a lack of understanding of invertebrate behaviors does not preclude their capacity for sentience or their ability to respond to negative experiences in a non-anthropocentric manner that could cause suffering [12].

Concurrent with this growing ethical awareness, significant efforts have been made in experimental research to improve invertebrate welfare [5,7]. However, relevant information about the implications for invertebrate welfare is dispersed, limited, and sometimes contradictory. As a result, the current study conducts a comprehensive review of the current landscape of animal ethics and echinoderm welfare, with the goal of contributing to the development of a comprehensive framework for invertebrate welfare.

2. Ethical Concerns and the 5Rs Principle

In general, ethics is concerned with situations involving various types of conflicts. While there are many different types of conflicts in science, not all of them require ethical consideration. The most widely accepted criterion for determining the relevance of ethical reflection is whether or not it has an impact on humans [13]. This viewpoint clearly states that all human beings are morally concerned (an anthropocentric doctrine). However, the same cannot be said for non-anthropocentric living beings. As a result, determining which living beings should be regarded as ethical considerations becomes difficult.

One approach is to use humans as a reference point to determine whether these living beings meet criteria for sentience or consciousness, which include behavioral, evolutionary, and physiological considerations [14]. Scientific evidence supports arguments about which living beings should be considered ethically [6,12,15–17]. With our improved understanding of animal suffering, the scope of moral consideration has expanded to include all vertebrates, as well as octopuses as the sole group of invertebrates. Nonetheless, the ethical discourse is increasingly shifting toward considering animal species regardless of scientific knowledge about their capacity for suffering [18]. Significant efforts have recently been directed toward developing an ethical framework that takes invertebrates into account. Significant progress has been made in this regard for two distinct groups, namely crustaceans [19] and echinoderms [8,20].

The 3Rs principle, proposed by Russell and Burch (1959) [21], is the globally accepted tool that currently guides animal research practices. Replacement is one aspect of this principle that advocates for the use of scientific methods involving non-sentient material to replace those involving conscious living vertebrates [21]. Replacement should ideally encourage the use of lower levels of organization, such as cell culture, and even artificial models, such as computational simulations. However, because invertebrates are generally considered non-sentient animals, the replacement dimension tends to favor them as common models for experimental research. In this regard, the principle falls short in considering the inclusion of those invertebrates whose capacity for sentience is still debated. The 3Rs principle is built on a strong assumption: that non-human animals lower on the zoological scale lack sentience [22]. This assumption presents a significant limitation, especially considering the vast number and diversity of invertebrates used in scientific research. Despite the widespread acceptance of the 3Rs as a policy tool aimed at alleviating animal suffering and reducing their use in research, its effectiveness in achieving these goals has been notably criticized [23]. In fact, the number of animals used for research in the European Union (EU) is now similar to that in the 1980s [24]. The three R principle was originally intended to establish ethical guidelines for animal experimentation, albeit with species-specific limitations. However, it did not address the epistemological issues raised via animal manipulation adequately [25]. This principle was later expanded to the 5Rs principle (Figure 1) to include two additional concepts emphasizing personal commitment: respect (establishing a respectful relationship with any living being regardless of its complexity or the knowledge we have of that living being) and responsibility (personal commitment of researchers to apply ethics concepts conscientiously) [8].

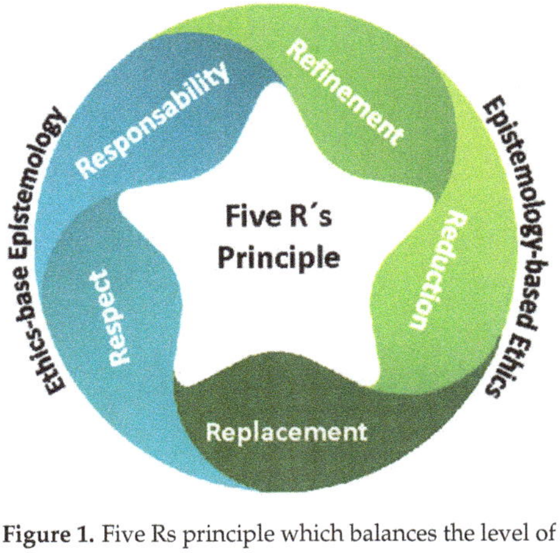

Figure 1. Five Rs principle which balances the level of importance of the ethical and epistemological approaches [8].

3. Ethical Approach in Echinoderms

3.1. Phylum Echinodermata

Echinoderms are a phylum of marine invertebrates. They are often key, long-living species that shape and maintain the status of many marine ecosystems, inhabiting a wide range of ecological niches from the abyssal depths of the oceans to the intertidal zone. Echinoderms, distinguished by their pentamerous radial body arrangement, form a monophyletic group with hemichordates, also known as acorn worms. These organisms exhibit significant diversity and widespread distribution across a variety of marine habitats, playing critical ecological roles in each setting [26]. The fundamental characteristics of this phylum have exhibited remarkable consistency since the Ordovician epoch (approximately 495–440 million years ago) [27], with an approximate enumeration of 7000 species [28].

The extant echinoderm assemblage includes approximately 7000 species divided into five distinct taxonomic groups: Asteroidea, which includes starfishes; Echinoidea, which includes sea urchins, sand dollars, and sea biscuits; Crinoidea, which includes sea lilies and feather stars; Ophiuroidea, which includes basket stars and brittle stars; and Holothuroidea, which includes sea cucumbers. This collective lineage has an exceptionally notable fossil record dating back to the Cambrian period, providing a solid foundation for comparative molecular studies spanning a wide range of meticulously documented divergence intervals [27,28] (Figure 2).

Figure 2. Example of classes of echinoderms: (**A**) *Cycethra verrucosa*, Asteroidea from South Atlantic; (**B**) *Arbacia dufresnii*, Echinoidea from South Atlantic; (**C**) *Ctenantedon kinziei*, Crinoidea from the Caribbean; (**D**) *Ophiocoma echinata*, Ophiuroidea from the Caribbean; (**E**) *Pearsonothuria graeffei*, Holothuroidea from the Indo-Pacific.

Echinoderms have a number of distinguishing characteristics that set them apart from their zoological counterparts. Notably, among these characteristics is a hydraulic water vascular system that is intricately linked with a distinct arrangement of a calcium carbonate endoskeleton known as stereom. Most echinoderms' developmental trajectories begin with larval stages, where intricate metamorphic processes culminate in the eventual manifestation of the adult form. These organisms exhibit a diverse range of life history traits, most notably sexual reproduction, though asexual reproduction is also observed. Echinoderm larvae are primitive, free-living, and planktonic in nature, with a wide range of morphology and functional characteristics that occasionally resemble those observed in hemichordate larvae. Following the metamorphic stage, the majority of mature echinoderms adopt a benthic lifestyle with radial symmetry and a typical pentameric structural composition. The organisms' internal architecture is intricately organized, with a reliance on calcium carbonate ossicles reinforced by a complex network of collagenous ligaments. Notably, any skeletal structures found in larvae adopt intricate rod-like configurations with distinct origins in Ophiuroidea and Echinoidea, whereas such features are absent in the larvae of the remaining three echinoderm classes. Echinoderms have distinct pentaradial symmetry in their adult forms. Nonetheless, their developmental origins can be traced back to bilaterally symmetrical larvae, a shared feature of the deuterostome clade (shown in Figure 2A). These distinguishing characteristics place echinoderms within an intriguing evolutionary framework, identifying them as invertebrate deuterostomes inextricably linked to vertebrate organisms [29].

Echinoderm species exhibit a variety of developmental strategies, ranging from direct development from a fertilized egg to an adult to indirect development, in which adults emerge from the metamorphosis of a larva with no relation to the adult. The cell structure and count of indirect developing species' long-living, feeding, bilaterally symmetrical larvae are very simple. There are numerous intermediate developmental stages, including facultative larval feeding and non-feeding larvae. Indirect development is primitive in echinoderms, and all five surviving classes, as well as the sister phylum Hemichordata, have dipleurula-type larvae [30].

Crinoids (Crinoidea) are sessile or free-living benthic animals with macrophagous filter-feeding habits. Some comatulids feed during the day, while others feed by extending the tips of their arms or moving only at night to avoid predators. Some forms prefer and actively seek out areas with flowing water for feeding (rheophilic), while others do not (rheophobic). They can be found in a variety of habitats, including as attachments to substrates via their arms or cirri, within caves or small crevices, beneath rocks, or as attachments via cirri to other invertebrates such as corals (epizoic). Crinoids are important in developmental biology because they are the only echinoderms that have a primitive tripartite coelom. They are significant in paleontology because they evolved during the Cambrian period and were a dominant and diverse component of Paleozoic benthic fauna [31].

Sea cucumbers (Holothuroidea) are mostly found in benthic marine habitats ranging from the littoral zone to the abyss. They are found in all oceans, with particular abundance in coral reefs. Sea cucumbers are primarily detritivores, feeding on both suspended particles and organic matter bound in sediment. The distribution of sea cucumbers is influenced by the type of substrate, as well as other environmental variables such as dominant currents, temperature, water salinity, and depth. Sea cucumbers' ecological importance is linked to their bioturbation activities, which involve the movement of organic material within sediments as well as the transfer of energy and materials at the water-sediment interface. The commercialization of sea cucumbers' body walls, also known as beche de mer or trepang, is economically significant. Out of the over a thousand known species, the fishing industry primarily focuses on around thirty. This practice has a long history in Chinese and Japanese culinary and medical traditions. Larger species are especially valuable. Trepang curing is applied to the body wall, resulting in a product with high nutritional value due to its high protein and low fat content. Furthermore, sea cucumbers contain biologically active compounds that are used to treat a variety of medical conditions, including HIV, cancer, and osteoarthritis [32,33].

Echinoids (Echinoidea) are found in a wide variety of geographical and bathymetric environments, from intertidal zones to depths of 5000 m. Regular sea urchins live on a variety of substrates, with the majority of them living on rocky or mobile substrates. Sand dollars and heart urchins, for example, can only be found on soft bottoms and frequently bury themselves. Many morphological differences between regular and irregular echinoids are caused by differences in lifestyle and feeding habits. Some populations of regular sea urchins exposed to wave action have developed digging behaviors to bury themselves slightly, whereas species not exposed to such conditions typically exhibit cryptic behaviors [34]. Sea urchins, as well as sea cucumbers, are valuable species, and the commercialization of sea urchins' gonads, also known as roe, is economically significant around the world [34,35].

Ophiuroids (Ophiuroidea) have a wide geographic distribution and a benthic lifestyle, having adapted to live in a variety of environments. They have been discovered in submarine hydrothermal vents and on bottoms with cold methane seeps, as well as in intertidal zones and abyssal regions from the tropics to the poles. The majority of species are usually found on soft substrates. They can be carnivorous, scavengers, sediment-consuming, or filter feeders. The majority use multiple feeding methods, though one usually takes precedence over the others. The majority are carnivorous, eating polychaetes, mollusks, and small crustaceans. Because of their diverse feeding strategies, they play an important role in trophic chains. They are not commercially important, unlike some other echinoderm groups. Nonetheless, due to their abundance, they can be used as environmental indicators [36].

Sea stars (Asteroidea) have a wide geographical range and a benthic habitat, inhabiting a variety of marine substrates where they can be abundant and visible. The vast majority are scavengers or opportunistic predators. Many species have generalist feeding habits and play an important role in the structure and functioning of marine communities as apex predators. They play an ecologically important role across latitudes by occupying different levels of the trophic chain, particularly as apex predators in rocky and coral reef ecosystems. This group's ecological success can be attributed to a variety of morphological and life history characteristics, including indeterminate growth, extraoral and intraoral digestion (providing access to a diverse diet), rapid prey detection and response, and the ability to anchor themselves to substrates using ambulacral tube feet [36,37].

3.2. *Historical Use of Echinoderms*

Because sea urchins have been consumed by humans throughout history, the human understanding of this Phylum dates back to prehistoric times [38]. They are depicted in "frescos cretenses", which date back 4000 years. For a long time, Eastern cultures have consumed and used sea cucumbers for medicinal purposes [39]. The earliest known of these animals dates from the period of Aristotle, who described the first known echinoderm in 350 B.C.E., over 2000 years ago. He described the feeding apparatus of sea urchins in his work "Historia Animalium", which is now known as *Aristotle's Lantern*. It is worth noting that Aristotle classified echinoderms as ostracoderms. The scholars of the time rekindled their interest in nature and began studying these creatures again. Klein coined the term "Echinodermata" in his 1734 work "Naturalis dispositio echinodermatum". However, he only used it to refer to sea urchins, not all the classes that are now known. Linnaeus classified the genera Asterias, Echinus, and Holothuria as Mollusca in the 10th edition of *Systema Naturae* (1758). The term "Echinodermata" resurfaced in 1792, when it was recognized that these animals were a distinct group of invertebrates, though sea cucumbers were not included. Later, Lamarck (1809) [40] grouped Echinodermata with the true Coelenterata in the Radiata group of invertebrates. It took nearly four decades for De Tornos (1839) [41] and Salacroux (1840) [42] to coin the term "Echinodermata". Due to its more advanced structural characteristics, it was argued in 1854 that Echinodermata did not belong with Coelelenterata. Echinodermata have since been recognized as a distinct clade of invertebrates.

Since the nineteenth century, the description of Echinodermata species has been a dominant focus in the literature. Initially, the focus was on species found along Europe's coasts, as evidenced by Frey and Leuckart's work in 1847 [43]. As a result of numerous oceanographic expeditions, this trend has expanded to include species from all over the world. While it is impractical to list every expedition and paper that resulted from it, some of the pioneering ones are worth mentioning. HMS Challenger and the Albatross laid the groundwork for hundreds of subsequent expeditions around the world. These expeditions have recently expanded to include deep-sea species as well as those found in intertidal and shallow waters [44–48]. Given that echinoderms have been collected worldwide for over 300 years to describe species, determine distributions, and populate museum collections, it is clear that hundreds of thousands of echinoderms were collected and preserved without much ethical thought.

3.3. *Echinoderms as Models*

Since the widespread use of the microscope in scientific research, echinoderms have served as experimental models. Due to the ease with which gametes could be obtained and the optical transparency of sea urchin embryos, they became valuable animal models in the mid-nineteenth century. Dufossé (1847) [49] and Derbès (1847) [50] provided early insights into fertilization and sea urchin embryo development via metamorphosis. Despite the fact that their work was overlooked in the scientific literature [51], sea urchin embryos remained useful as model organisms. Hertwig (1876) [52] made a pivotal discovery by demonstrating how sperm entered the female gamete, leading to embryo formation in the sea urchin *Toxopneustes lividus*. This discovery established sea urchin embryos as the gold standard for embryonic research, advancing our understanding of fertilization mechanisms, egg activation, cleavage, gastrulation, and early embryonic differentiation. They have also been useful in research on nervous system development, evolutionary development, and regeneration [51,53–56] (http://www.echinobase.org, accessed on 10 October 2023).

Sea stars were also recognized as important model organisms in the early twentieth century. Metchnikoff (1893) [57] discovered a cellular immune response in the bipinnaria larvae of sea stars, observing the rapid migration of mesenchymal cells to injury sites. He also discovered amoeboid cells' phagocytic activities [57]. Despite these groundbreaking discoveries, echinoderm larvae were not widely used as model organisms in immunology until the twenty-first century (Figure 2A) [58–60].

One of the most remarkable characteristics of echinoderms is their ability to regenerate lost structures, such as the arms of sea stars. In recent years, there has been a surge in research into the regenerative capacity of the echinoderm nervous system (NS). Studies have shown that holothurians can regenerate their radial nerve cord (RNC) after it has been transected, with the regenerated cords displaying a similar structure and function to that of the originals and without scar tissue formation—a common issue in vertebrate CNS regenerative responses. Follow-up studies have revealed that radial glia play an important role in the regenerative process, both in forming the bridge that connects the severed ends of the RNC and in generating new neurons and glia in the regenerated structure [61,62]. Several laboratories are currently working to identify the genes required for NS regeneration [63–67].

The capacity for evisceration and subsequent regeneration in sea cucumbers, as well as brittle stars and sea stars due to their ability to regrow arms, has positioned them as invaluable models in regeneration research [66,68–81]. With recent technological advancements and the availability of new tools, echinoderms have become outstanding model organisms for both scientific research and educational purposes, and in most cases individuals are handled alive [82,83] (www.Echinobase.org, accessed on 10 October 2023).

3.4. Echinoderms Nervous System

In echinoderms, the nervous system is organized in accordance with the general pentameric pattern of the body plan. Each radius has its own radial nerve cord, which runs the length of the proximo-distal axis and terminates at the distal tip of the arm (in stellate echinoderms) or near the aboral pole (in globose forms). A circumoral nerve ring joins all five individual radial nerve cords at the body's oral pole to form a single anatomical entity. The radial nerve cords and the nerve ring comprise the echinoderm's central nervous system (CNS) (Figure 3). This CNS is an anatomically and histologically distinct agglomeration of neurons and glial cells associated with an extensive neuropil (densely interwoven neuronal processes) found nowhere else in the body and is in charge of the initiation and coordination of various body-wide responses [84].

Adult echinoderm neuroanatomy is distinguished by the presence of distinct superimposed domains or layers of nervous tissue located at different levels relative to the oral–aboral axis. These domains are known as the ectoneural and hyponeural systems [85–87]. The ectoneural system is located around the mouth, either within or directly beneath the oral epidermis. It is always present both in the nerve ring and radial nerve cords of all echinoderms, shows the most consistent organization across the phylum, and is the predominant part of the nervous system in all classes except crinoids. The hyponeural system may or may not be a part of the nerve ring or radial nerve cords. Its organization differs between classes and is generally related to the degree of development of large muscles. When hyponeural tissue is present, it acts as a second (usually thinner) layer of nervous tissue that directly overlies the aboral surface of the respective ectoneural cords [87,88].

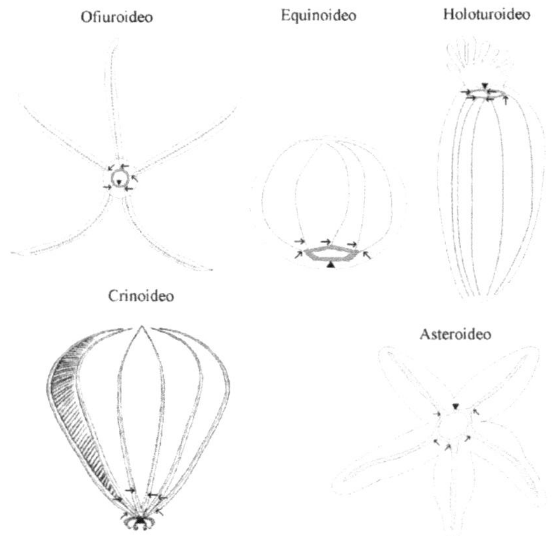

Figure 3. Schematic representation of the central nervous system of the phylum Echinodermata. The arrows indicate the radial nerves cords and the point of the arrows indicates the circumoral ring.

This simplified classification of the echinoderm nervous system, however, does not fully account for recent reports from the last decade, revealing unexpected nervous system elements that do not neatly align with the three aforementioned divisions [64,84]. The nervous system (NS) of echinoderms is one of the most fascinating aspects. Given their close evolutionary relationship with chordates, as well as their radial symmetry and lack of prominent ganglia or centralized nervous structures associated with cephalization found in most other animals, the echinoderm nervous system has been regarded as pivotal to understanding the evolution of the chordate nervous system [29,84]. In fact, it has been suggested that the centralized nervous system is a plesiomorphic (ancestral) condition in echinoderms and may also be a plesiomorphic trait at the level of the Deuterostomia [84].

In the nervous system, including the ectoneural and hyponeural subsystems, tissue is organized as a neuroepithelium made up of two major cell types: radial glial cells and neurons. The cell types have a similar relative abundance in the radial nerve cord's ectoneural and hyponeural bands, with radial glial cells accounting for 60–70% of the total cell population. On the other hand, neuronal cells are more abundant in the circumoral nerve ring, accounting for only 45% of the cell composition formed by glia [89]. Echinoderm glial cells share significant morphological similarities with the radial glia of chordates, including the orthogonal orientation of the cell's main axis to the plane of the neuroepithelium, the presence of long thick bundles of intermediate filaments, and the presence of short protrusions branching off at a right angle from the main processes and penetrating into the surrounding neural parenchyma. The cell bodies of the majority of radial glial cells in echinoderms are located at the apical surface of the neuroepithelium. Some of them, however, are bipolar, with the apical and basal processes extending from opposite poles of cell bodies located at different depths within the neural parenchyma. Radial glial cells are the most common type of glial cell in echinoderms (though they are unlikely to be perfectly homogeneous), and they perform a variety of functions. In addition to the radial glia of the CNS, other glial cell types associated with the peripheral nervous system may exist [84]. Radial glia have more morphological similarities than do chordates because they are the main proliferative population in nervous tissue and thus capable of giving rise to new neurons, both in non-injured and regenerating CNSs. There are, however, differences between radial glia in echinoderms and chordates. Another significant difference is that radial glia are the only major glial cell type in the adult CNS of echinoderms, whereas in higher vertebrates, radial glia predominates in embryogenesis but then mostly disappear from the mature nervous system by giving rise to a plethora of more specialized cell types [90,91]. Radial glia are common in the adult CNS of lower vertebrates, but they frequently co-exist with other abundant specialized glia, such as those of oligodendrocytes [92,93].

The neural parenchyma of the CNS is made up of neurons with somata and neurites. The most common neuronal morphology ranges from unipolar and bipolar to multipolar. In echinoderms, neurons can be classified quite easily by size as normal neurons, which are dominant in most classes, and the giant neurons of ophiuroids. The first class has small somata (about 5 m in diameter) that produce very thin give processes (0.1–1 m in cross-section) with numerous local swellings (varicosities) along their length. Because they are typically found near or embedded within calcareous structures, they are less appealing to neurobiologists, particularly electrophysiologists who study their electrical properties. As a result, while other animal groups became popular neurobiology research subjects, echinoderms were largely overlooked for a long time. Even today, there are surprisingly few electrophysiological studies on echinoderms [94,95].

Communication between neurons in traditional chemical synapses was previously thought to be absent in echinoderms [85,96]. However, this long-held belief may be due to a lack of adequate tools for dealing with the difficulties imposed by the endoskeleton. Since the optimization of sample preparation protocols, there has been evidence demonstrating the presence of typical chemical synapses that occur on a regular basis in the CNS of Echinodermata [89,97]. These findings are consistent with those found in the sea urchin genome, where the genes required for synapse formation were discovered [83]. There are also different types of synapses: unsheathing synapses (with the pre-synaptic terminal wrapped around the post-synaptic process), passant synapses between parallel nerve fibers, and complex synapses with a pre-synaptic terminal forming two or more synapses in different post-synaptic processes, or, conversely, a single post-synaptic neuron receiving synaptic input from multiple pre-synaptic axons [97]. Echinoderm CNSs have regional differences in cellular composition as well as a complex internal spatial segregation of different cell types. The radial nerve cord is made up of repetitive units [85,98,99].

The CNS of echinoderms generates complex, coordinated, and directional behavioral responses to various sensory stimuli. Although the molecular and cellular mechanisms underlying these behaviors remain unknown, it is known that an echinoderm's CNS contains a large number of neurotransmitters from all major groups, including acetylcholine, aminoacids, monoamines, neuropeptides, and gases. Acetylcholine appears to mediate muscle contraction due to its function as a major excitatory neurotransmitter [88,100,101]. Post-synaptic nicotinic and muscarinic receptors have also been identified, and acetylcholinesterase (the enzyme required to hydrolyze acetylcholine at synapses) activity has been detected in both ectoneural and hyponeural systems [85,102]. GABAergic neurons proliferate throughout the CNS, including the radial nerve cord, nerve ring, and podial nerves, as well as the nerves and visceral plexi [84,103]. GABA is involved in both echinoderm muscle contraction and relaxation, depending on the post-synaptic receptor (GABA A or GABA B) present in neuromuscular junctions [100]. L-glutamate is and excitatory neurotransmitter in the ectoneural subsystem of echinoderms [104,105]. L-glutamate is also a neurotransmitter capable of eliciting the arm autotomy response, whereas acetylcholine acts as an antagonist of L-glutamate [104]. The research on serotonin as a neurotransmitter in the adult CNS is limited. There have been reports of its presence in muscles and basiepithelial plexi [106–108]. Furthermore, neuropharmacological studies have revealed that serotonin regulates muscular contraction by inhibiting the excitatory effect of acetylcholine [107] and may also be involved in the regulation of post-traumatic regeneration [108]. Catecholamines such as dopamine and noradrenaline have been found in the ectoneural only and appear to be involved in the movement of tube feet [109,110] and to be fundamental for the righting response of the sea urchin [111]. Histamine data of the sea urchin are extremely limited; they have only been studied in one species of cucumber, and the latter appears to be involved in sensory systems, as it was found in tentacles and body wall papillae, and to project its axons directly to the nerve ring and the radial nerve cord [112].

The enzyme nitric oxide synthase, which produces NO, was discovered in both the ectoneural and hyponeural parts of the radial nerve cord of adult sea stars, as well as in some radial glial cells [113,114]. Apparently, NO is involved in the relaxation of viscera and tube feet [115,116].

In addition to these phylogenically widespread neurotransmitters, echinocherms contain specific neuropeptides from the SALMFamide family [88,98,116–119]. These neuropeptides relax the visceral musculature as well as the muscles of the body wall [116,120].

All of the information above suggests that the echinoderm CNS is more complex than previously thought and that, despite the lack of a centralized brain, it is possible for it to elicit complex individual and social behavior. Recent advances in knowledge of this group have provided new insights suggesting that echinoderms are sentient animals capable of suffering pain.

3.5. Pain and Echinoderms

Even though the debate over pain perception in invertebrates is still ongoing, it is critical to recognize that the absence of evidence of painful sensations should not be interpreted as conclusive proof of pain absence in this group. As scientists, we are responsible for treating the organisms we study with dignity and ensuring their wellbeing. To accomplish this, we must first understand the concepts of analgesia, sedation, and anesthesia. It is also critical to understand how to correctly administer anesthesia, which response variables to consider, and which substances to use.

Various agents have been used to anesthetize echinoderms [121]. Iso-osmotic solutions such as $MgCl_2$, $MgSO_4$, or Ca^{2+}-free seawater are common echinoderm anesthetics. The mechanism of action of these agents is to destabilize membrane potential, preventing pain signals from propagating. Additionally, local anesthetics that block neural stimulation of the muscle have been employed when necessary. The local anesthetics MS222 [119,122–125] and propylene phenoxetol [122,126–131] have both demonstrated effectiveness in studies involving echinoderm connective tissue.

MS-222 (IUPAC name 3-amino benzoic acid, ethyl ester, and methanesulfonate salt, also known as ethyl m-aminobenzoate or tricaine methanesulfonate) is a local anesthetic of the ester type. Its structure is similar to that of other local anesthetics, such as benzocaine, implying that it likely functions similarly by impeding axonal conduction via interference with membrane depolarization [132]. Originally developed as a fish anesthetic [133], it has since been widely used on a variety of invertebrates (National Research Council, 1981).

In ophiuroids, "propylene phenoxetol" is thought to act as a local anesthetic that inhibits axonal conduction [131]. Because this volatile liquid is not water-soluble, maintaining known concentrations of the compound in a medium is difficult. Furthermore, there have been some doubts about the precise identity of this compound. While certain compounds possessing anesthetic effects in echinoderms have been studied, the practice of using them during individual manipulations is not yet standardized because there is neither awareness nor a normative mandate for it.

4. Echinoderm Welfare

Animal welfare is now recognized as a scientific discipline that includes ethology, physiology, pathology, biochemistry, genetics, immunology, nutrition, cognitive–neural studies, veterinary care, and ethics [134–140]. While the assessment of animal welfare has traditionally focused on vertebrates [141], the vast diversity of invertebrates presents a significant challenge.

Firstly, the fundamental indicators (such as cortisol levels, longevity, feeding rate, and behavior) utilized in welfare assessments follow a reductionist approach. Secondly, this assessment only encompasses two of the "Three Conceptions" (basic health and function, and natural living) and four of the "Five Domains" (nutrition, environment, health, and behavior) that should ideally be evaluated. By definition, this basic assessment remains incomplete. Moreover, in invertebrates, except for cephalopods, evaluating the mental state domain and the conception of affective state is currently extremely challenging. Although we lack the tools to assess invertebrate mental states, there is undeniable evidence of social behavior in many species, implying that a lack of interactions with these organisms may have a negative impact on their mental wellbeing [142]. Even if a comprehensive welfare assessment for invertebrates is not currently feasible, we must ensure that the best welfare conditions are met.

A comprehensive assessment of animal welfare, according to Botreau et al. (2007) [143], requires a well-defined set of criteria. The following guidelines should be followed when developing these criteria:

- Each and every significant aspect must be addressed in order for the assessment to be comprehensive;
- The criteria must not be redundant or irrelevant;
- Each criterion must be independent of the others;
- All stakeholders must agree on the criteria, and they must have a practical basis;
- The criteria, as well as their application, should be transparent and simple to understand;
- The number of criteria should be limited to 12 at most.

Given the diversity of invertebrates, it is critical to recognize the importance of developing a specific set of criteria for assessing the welfare of each phylum or even each order within this group. There are several indicators that provide information about an animal's wellbeing in response to its experiences. These indicators are based on a thorough understanding of how individuals respond physiologically and behaviorally to various conditions. The interpretation of these indicators is dependent on one's attitude toward animal welfare. Exploration, hunting or foraging, socialization, parental care, play, and sexual activity are all expected to develop as inherent behavioral elements [144,145]. As a result, when an animal is kept in captivity, it is critical to understand the specific fundamental characteristics of each species that exist in their natural environment in order to recreate them. This meets the need to "explore, solve problems, and overcome challenges" [146] (p. 623).

In order to achieve the application of echinoderm wellbeing, the following non-invasive echinoderm assessments have recently been developed as indicators for use in the aquaculture industry and research, providing valuable insights into individual wellbeing.

- Tube feet adhesion: Echinoderms move by using their tube feet, which are frequently adhesive to the bottom or walls of an aquarium in captivity. A detached or loosely attached echinoderm indicates that the individual is not in good health.
- Echinoderms have distinct righting behavior because they are most vulnerable with their oral face above [147,148]. The speed at which this behavior is carried out can be used to gauge their physiological activity, health, and overall condition. This indicator must first be established for the species, and size and sex must be evaluated before establishing a normal time of righting for each species. After you have set a time, you can use it as a stress indicator.
- Echinoderms use their spines/pedicelaria/arms for a variety of functions (feeding, moving, defense, and eating), so their response to stimuli is a good indicator of their health. An echinoderm that is nonresponsive or responds slowly indicates that the individual is not in good health.
- Feeding behavior: like any other animal's, the feeding behavior of echinoderms is an indicator of their health. Echinoderms feeding and defecating indicate healthy individuals.
- Epidermis appearance: A healthy echinoderm has a shiny, non-interrupted epidermis. In contrast, the presence of reddish or blackish coloration, as well as inflammation and mucus, indicates the presence of an infection.

5. Conclusions

The majority of ethical and welfare approaches in animal research have primarily focused on vertebrates. Addressing invertebrate welfare presents unique challenges, and researchers will require time to fully integrate these concepts. However, significant progress in this area has been made. With rising public awareness and concern, there is a chance that ethical concepts will be adopted more quickly. In the near future, this could lead to the establishment of guidelines, norms, and laws in this domain.

In recent years, it has become clear that advanced invertebrates have qualities such as self-awareness and sentience, as well as the ability to experience pain, though precisely defining and understanding it in their context is difficult. Animal welfare legislation in various countries supports this viewpoint. While many invertebrates have learning and memory abilities, there are significant structural and physiological differences between animal groups. Some researchers believe that these differences indicate that, despite their self-protective behaviors, advanced invertebrates are incapable of feeling pain. Regardless of the validity of this argument, it is critical for humanity, particularly scientists, to take invertebrate welfare seriously and to treat them with care, both in captivity and in their natural habitats. The scientific community has responded by advocating for invertebrate welfare to be considered in breeding or holding facilities, laboratories, and under natural conditions whenever possible. In addition, if advanced invertebrate experimentation is deemed necessary, appropriate anesthesia methods should be considered.

This research provides evidence of sentience within the echinoderm group, highlighting important advances in our understanding of their biology and physiology. However, there are still notable gaps in information that require further investigation. Regardless of the current state of knowledge regarding sentience in echinoderms, there is a growing moral consideration to implement practices ensuring animal welfare when utilizing individuals in research. The 5Rs principle provides a useful conceptual framework in this situation by highlighting the importance of Replacement, reduction, and refinement as well as respect and responsibility in researchers' practices while working with echinoderms.

Author Contributions: Both authors, A.C.C.-A. and T.R., contributed to conceptualization, and writing and editing the manuscript. All authors have read and agreed to the published version of the manuscript.

Funding: This research was funded by CONICET, grant number PIP 11220210100013CO.

Institutional Review Board Statement: Not applicable.

Informed Consent Statement: Not applicable.

Data Availability Statement: Not applicable.

Acknowledgments: We acknowledge the three anonymous reviewers whose valuable suggestions significantly contributed to the improvement of the manuscript.

Conflicts of Interest: The authors declare no conflict of interest.

References

1. Kellert, S.R. Values and Perceptions of Invertebrates. *Conserv. Biol.* **1993**, *7*, 845–855. [CrossRef]
2. Schmidt-Rhaesa, A.; Schwartz, M. Nematomorpha from the Philippines, with Description of Two New Species. *Zootaxa* **2016**, *4158*, 246–260. [CrossRef]
3. Pollo, S.; Vitale, A. Invertebrates and Humans: Science, Ethics, and Policy. In *The Welfare of Invertebrate Animals*; Carere, C., Mather, J., Eds.; Springer: Berlin/Heidelberg, Germany, 2019; pp. 7–22.
4. Crespi-Abril, A.C.; Rubilar, T. Ética e Invertebrados: Análisis de los Casos de los Cefalópodos y Equinodermos. *Rev. Latinoam. Estud. Crít. Anim.* **2018**, *8*, 210–232.
5. Horvath, K.; Angeletti, D.; Nascetti, G.; Carere, C. Invertebrate Welfare: An Overlooked Issue. *Ann. Dell'istituto Super. Di Sanità* **2013**, *49*, 9–17.
6. Mather, J.A. An Invertebrate Perspective on Pain. Anim. *Sentience* **2016**, *1*, 12. [CrossRef]
7. Carere, C.; Mather, J. *The Welfare of Invertebrate Animals*; Springer: Berlin/Heidelberg, Germany, 2019. [CrossRef]
8. Crespi-Abril, A.C.; Rubilar, T. Moving Forward in the Ethical Consideration of Invertebrates in Experimentation: Beyond the Three R's Principle. *Rev. Biol. Trop.* **2021**, *69* (Suppl. S1), 346–357. [CrossRef]
9. Mather, J.A. Why (and How) Personalities in Invertebrates? *Curr. Zool.* **2012**, *58*, 566. [CrossRef]
10. Schmidt-Rhaesa, A.; Harzsch, S.; Purschke, G. *Structure and Evolution of Invertebrate Nervous Systems*; Oxford University Press: Oxford, UK, 2015.

11. Mikhalevich, I.; Powell, R. Minds without Spines: Evolutionarily Inclusive Animal Ethics. *Anim. Sentience* **2020**, *10*, 1–25. [CrossRef]
12. Ellwood, R.W. Pain and Suffering in Invertebrates? *ILAR J.* **2011**, *52*, 175–184. [CrossRef]
13. Guariglia, O.; Vidiella, G. *Breviario de ética*; Edhasa: Buenos Aires, Argentina, 2011.
14. Edelman, D.B.; Seth, A.K. Animal Consciousness: A Synthetic Approach. *Trends Neurosci.* **2009**, *32*, 476–484. [CrossRef]
15. Chandroo, K.P.; Duncan, I.J.H.; Moccia, R.D. Can Fish Suffer? Perspectives on Sentience, Pain, Fear, and Stress. *Appl. Anim. Behav. Sci.* **2004**, *86*, 225–250. [CrossRef]
16. Griffin, D.R.; Speck, G.B. New Evidence of Animal Consciousness. *Anim. Cognit.* **2004**, *7*, 5–18. [CrossRef]
17. Lewbart, G.A.; Mosley, C. Clinical Anesthesia and Analgesia in Invertebrates. *J. Exot. Pet Med.* **2012**, *21*, 59–70. [CrossRef]
18. Birch, J. Animal Sentience and the Precautionary Principle. *Anim. Sentience* **2017**, *16*, 1. [CrossRef]
19. Birch, J.; Burn, C.; Schnell, A.; Browning, H.; Crump, A. *Review of the Evidence of Sentience in Cephalopod Molluscs and Decapod Crustaceans*; LSE: London, UK, 2021.
20. Rubilar, C.T.; Crespi-Abril, A.C. Does Echinoderm Research Deserve an Ethical Consideration? *Int. J. Trop. Biol.* **2017**, *65*, S11–S22. [CrossRef]
21. Russell, W.M.S.; Burch, R.L. *The Principles of Humane Experimental Technique*; Methuen: London, UK, 1959.
22. Tomasik, B. Suffering in Animals vs. Humans, Essays on Reducing Suffering. Available online: http://reducing-suffering.org/suffering-in-animals-vs-humans/ (accessed on 10 October 2023).
23. Blattner, C.E. Rethinking the 3Rs: From Whitewashing to Rights. In *Animal Experimentation. Working Towards a Paradigm Change*; Hermann, K., Jayne, K., Eds.; Brill: Amsterdam, The Netherlands, 2019; pp. 168–193.
24. Taylor, K.; Rego, L. EU Statistics on Animal Experiments for 2014. *ALTEX* **2016**, *33*, 465–468. [CrossRef] [PubMed]
25. Johnson, J.; Degeling, C. Animals-as-Patients: Improving the Practice of Animal Experimentation. *Between Species* **2012**, *15*, 43–58. [CrossRef]
26. Brusca, R.C.; Moore, W.; Schuster, M. *Invertebrates*; Sinauer Associated, Inc.: Sunderland, MA, USA, 2016.
27. Telford, M.J.; Lowe, C.J.; Cameron, C.B.; Ortega-Martinez, O.; Aronowicz, J.; Oliveri, P.; Copley, R.R. Phylogenomic Analysis of Echinoderm Class Relationships Supports Asterozoa. *Proc. R. Soc. B* **2014**, *281*, 20140479. [CrossRef] [PubMed]
28. Smith, A.B. Echinoderms (Other Than Echinoids). In *Encyclopedia of Geology*; Cocks, R., Plimer, I., Eds.; Elsevier: Oxford, UK, 2005; pp. 334–341.
29. Arnone, M.I.; Byrne, M.; Martinez, P. Echinodermata. In *Evolutionary Developmental Biology of Invertebrates 6*; Wanninger, A., Ed.; Springer: Vienna, Austria, 2015. [CrossRef]
30. Amemiya, C.T.; Miyake, T.; Rast, J.P. Echinoderms. *Curr. Biol.* **2005**, *15*, R944–R946. [CrossRef]
31. Hess, H.; Messing, C.; Ausich, W.I. *Treatise on Invertebrate Paleontology*; The University of Kansas Paleontological Institute: Lawrence, KS, USA, 2011.
32. Chen, J. Overview of Sea Cucumber Farming and Sea Ranching Practices in China. *SPC Beche-de-mer Inf. Bull.* **2003**, *18*, 18–23.
33. Slater, M.; Chen, J. Sea Cucumber Biology and Ecology. In *Echinoderm Aquaculture*; John Wiley & Sons: Hoboken, NJ, USA, 2015; pp. 47–55.
34. Lawrence, J.M. (Ed.) *Sea Urchins: Biology and Ecology*; Academic Press: Cambridge, MA, USA, 2013.
35. Sun, J.; Chiang, F.S. Use and Exploitation of Sea Urchins. In *Echinoderm Aquaculture*; John Wiley & Sons: Hoboken, NJ, USA, 2015; pp. 25–45.
36. López-Pérez, A.; Granja-Fernández, R.; Aparicio-Cid, C.A.; Zepeta-Vilchis, R.C.; Torres-Huerta, A.M.; Benítez-Villalobos, F.; López-López, D.A.; Cruz-Antonio, C.; Valencia-Méndez, O. Corales Pétreos, Equinodermos y Peces Asociados a Comunidades y Arrecifes Coralinos del Parque Nacional Huatulco, Pacífico Sur Mexicano. *Rev. Mex. Biodivers.* **2014**, *85*, 1145–1159. [CrossRef]
37. Luna Salguero, B.; Bonilla, H. Estructura Comunitaria y Trófica de las Estrellas de Mar (Echinodermata: Asteroidea) en Arrecifes Rocosos de Loreto, Golfo de California, México. *Hidrobiológica* **2010**, *20*, 127–134.
38. Lawrence, J.M. Sea Urchin Roe Cuisine. In *Developments in Aquaculture and Fisheries Science*; Elsevier: Amsterdam, The Netherlands, 2007; Volume 37, pp. 521–523.
39. Brown, N.; Eddy, S. *Echinoderm Aquaculture*; Wiley-Blackwell: Hoboken, NJ, USA, 2015.
40. Lamarck, J.B. *Philosophie Zoologique, ou, Exposition des Considérations Relative à l'Histoire Naturelle des Animaux*; Librairie F. Savy: Paris, France, 1809.
41. De Tornos, L. *Compendio de Historia Natural Dividido en los Tres Ramos de Mineralogía, Botánica y Zoología*; Imprenta Salvador Albert: Madrid, Spain, 1839.
42. Salacroux, A.P.G. *Nuevos Elementos de Historia Natural*; Imprenta de Verges: Madrid, Spain, 1840.
43. Frey, H.; Luckart, R. *Zootomie*; Voss: Leipzig, Germany, 1847.
44. Agassiz, A. *Report on the Echinoidea Dredged by H.M.S. Challenger during the Years 1873–1876. Zoology Part IX Challenger Reports*; Johnson Reprint Corporation: New York, NY, USA, 1881; Volume 3, pp. 1–321.
45. Théel, H. *Report on the Holothurioidea Dredged by H.M.S. Challenger during the Years 1873–1876. Zoology Part IX Challenger Reports*; Johnson Reprint Corporation: New York, NY, USA, 1882.
46. Agassiz, A. Notice of Calamocrinus Diomedeae, a New Stalked Crinoid from the Galapagos, Dredged by the U. S. Fish Commission Steamer 'Albatross'. *Bull. Museum Comp. Zool.* **1890**, *20*, 165–167.
47. Clark, H.L. Papers from the Hopkins Stanford Galapagos Expedition. *Proc. Wash. Acad. Sci.* **1902**, *4*, 521–531.

48. De Morgan, W. The Echinoderms Collected by the "Huxley" from the North Side of the Bay of Biscay in August, 1906. *J. Mar. Biol. Assoc.* **1913**, *9*, 530–541. [CrossRef]
49. Dufossé, A. Observations sur le Développement des Oursins. *Ann. Des Sci. Nat.* **1847**, *7*, 44–52.
50. Derbès, A.A. Observations sur le Méchanisme et les Phénomènes qui Accompagnent la Formation de l'Embryon chez l'Oursin Comestible. *Ann. Sci. Nat. Zool.* **1847**, *8*, 80–98.
51. Briggs, E.; Wessel, G.M. In the Beginning… Animal Fertilization and Sea Urchin Development. *Dev. Biol.* **2006**, *300*, 15–26. [CrossRef]
52. Hertwig, R. *Zur Histologie der Radiolarien: Untersuchungen über den Bau und die Entwicklung der Sphaerozoiden und Thalassicolliden*; W. Englemann: Leipzig, Germany, 1876.
53. Boveri, T. Die Polarität von Oocyte, Ei und Larve des Strongylocentrotus lividus. *Zool. Jahrb. Abt. Anat. Ontog. Tiere* **1901**, *14*, 630–653.
54. Dolmatov, I.Y. *Structure of the Aquapharyngeal Complex in the Holothurian Eupentacta Fraudatrix under the Normal Conditions and during Regeneration*; Canadian Science Dissertation; Institute of Marine Biology, Far East Division of the USSR Academy of Sciences: Vladivostok, Russia, 1988.
55. McClay, D.R. Evolutionary Crossroads in Developmental Biology: Sea Urchins. *Development* **2011**, *138*, 2639–2648. [CrossRef]
56. Garner, S.; Zysk, I.; Byrne, G.; Kramer, M.; Moller, M.; Taylor, V.; Burke, R.D. Neurogenesis in Sea Urchin Embryos and the Diversity of Deuterostome Neurogenic Mechanisms. *Development* **2016**, *143*, 286–297. [CrossRef] [PubMed]
57. Metchnikoff, I. *Lectures on the Comparative Pathology of Inflammation, Delivered at the Pasteur Institute in 1891*; Forgotten Books: London, UK, 1893.
58. Pinsino, A.; Thorndyke, M.C.; Matranga, V. Coelomocytes and Post-Traumatic Response in the Common Sea Star *Asterias rubens*. *Cell Stress Chaperones* **2007**, *12*, 331–341. [CrossRef] [PubMed]
59. Furukawa, R.; Funabashi, H.; Matsumoto, M.; Kaneko, H. Starfish ApDOCK Protein Essentially Functions in Larval Defense System Operated by Mesenchyme Cells. *Immunol. Cell Biol.* **2012**, *90*, 955–965. [CrossRef] [PubMed]
60. Ho, E.C.H. *A Simple Animal Model for Characterizing Gene Regulatory Control of an Immune Response*; University of Toronto (Canada): Toronto, ON, Canada, 2016.
61. Mashanov, V.S.; Zueva, O.R.; Garcia-Arraras, J.E. Radial Glial Cells Play a Key Role in Echinoderm Neural Regeneration. *BMC Biol.* **2013**, *11*, 49. [CrossRef]
62. Mashanov, V.; Zueva, O. Radial Glia in Echinoderms. *Dev. Neurobiol.* **2019**, *79*, 396–405. [CrossRef] [PubMed]
63. Ben Khadra, Y.; Said, K.; Thorndyke, M.; Martinez, P. Homeobox Genes Expressed during Echinoderm Arm Regeneration. *Biochem. Genet.* **2014**, *52*, 166–180. [CrossRef]
64. Díaz-Balzac, C.A.; García-Arrarás, J.E. Echinoderm Nervous System. In *Oxford Research Encyclopedia of Neuroscience*; Oxford University Press: Oxford, UK, 2018.
65. Dolmatov, I.Y. Molecular Aspects of Regeneration Mechanisms in Holothurians. *Genes* **2021**, *12*, 250. [CrossRef]
66. Medina-Feliciano, J.G.; García-Arrarás, J.E. Regeneration in Echinoderms: Molecular Advancements. *Front. Cell Dev. Biol.* **2021**, *9*, 768641. [CrossRef]
67. Mashanov, V.S.; Ademiluyi, S.; Machado, D.J.; Reid, R.; Janies, D. Echinoderm Radial Glia in Adult Cell Renewal, Indeterminate Growth, and Regeneration. *Front. Neural Circuits* **2023**, *17*, 1258370. [CrossRef]
68. Candia-Carnevali, M.D.; Bonasoro, F.; Lucca, E.; Thorndyke, M.C. Pattern of Cell Proliferation in the Early Stages of Arm Regeneration in the Feather Star *Antedon mediterranea*. *J. Exp. Zool.* **1995**, *272*, 464–474. [CrossRef]
69. Dolmatov, I.Y.; Ginanova, T.T. Muscle Regeneration in Holothurians. *Microsc. Res. Tech.* **2001**, *55*, 452–463. [CrossRef] [PubMed]
70. Candelaria, A.G.; Murray, M.; File, S.K.; García-Arrarás, J.E. Contribution of Mesenterial Muscle Dedifferentiation to Intestine Regeneration in the Sea Cucumber *Holothuria glaberrima*. *Cell Tissue Res.* **2006**, *325*, 55–65. [CrossRef] [PubMed]
71. San Miguel-Ruiz, J.E.; García-Arrarás, J.E. Common Cellular Events Occur During Wound Healing and Organ Regeneration in the Sea Cucumber *Holothuria glaberrima*. *BMC Dev. Biol.* **2007**, *7*, 115–124. [CrossRef]
72. Mashanov, V.S.; Zueva, O.R.; Heinzeller, T. Regeneration of the Radial Nerve Cord in a Holothrian: A Promising New Model System for Studying Post-Traumatic Recovery in the Adult Nervous System. *Tissue Cell* **2008**, *40*, 351–372. [CrossRef] [PubMed]
73. Carnevali, M.D.C.; Burighel, P. Regeneration in Echinoderms and Ascidians. eLS 2010. Available online: https://onlinelibrary.wiley.com/doi/abs/10.1002/9780470015902.a0022102 (accessed on 15 September 2023).
74. Mashanov, V.S.; Zueva, O.R.; García-Arrarás, J.E. Transcriptomic Changes During Regeneration of the Central Nervous System in an Echinoderm. *BMC Genom.* **2014**, *15*, 357. [CrossRef]
75. Mashanov, V.S.; Zueva, O.R.; García-Arrarás, J.E. Expression of Pluripotency Factors in Echinoderm Regeneration. *Cell Tissue Res.* **2015**, *359*, 521–536. [CrossRef]
76. Mashanov, V.S.; Zueva, O.R.; García-Arrarás, J.E. Inhibition of Cell Proliferation Does Not Slow Down Echinoderm Neural Regeneration. *Front. Zool.* **2017**, *14*, 12. [CrossRef]
77. Ferrario, C.; Ben Khadra, Y.; Czarkwiani, A.; Zakrzewski, A.; Martinez, P.; Colombo, G.; Bonasoro, F.; Carnevali, M.D.C.; Oliveri, P.; Sugni, M. Fundamental Aspects of Arm Repair Phase in Two Echinoderm Models. *Dev. Biol.* **2018**, *433*, 297–309. [CrossRef]
78. Dolmatov, I.Y. Variability of Regeneration Mechanisms in Echinoderms. *Russ. J. Mar. Biol.* **2020**, *46*, 391–404. [CrossRef]
79. Byrne, M. The Link Between Autotomy and CNS Regeneration: Echinoderms as Non-Model Species for Regenerative Biology. *BioEssays* **2020**, *42*, 1900219. [CrossRef]

80. Dolmatov, I.Y.; Kalacheva, N.V.; Mekhova, E.S.; Frolova, L.T. Autotomy and Regeneration of the Visceral Mass in Feather Stars. *Zoomorphology* **2020**, *139*, 171–187. [CrossRef]
81. Magalhães, F.; Andrade, C.; Simões, B.; Brigham, F.; Valente, R.; Martinez, P.; Rino, J.; Sugni, M.; Coelho, A.V. Regeneration of Starfish Radial Nerve Cord Restores Animal Mobility and Unveils a New Coelomocyte Population. *Cell Tissue Res.* **2023**, 1–16. [CrossRef]
82. Eisen, A. A Holistic Approach to Teaching a Laboratory Using Sea Urchin. Development as an Example System. In *Tested Studies for Laboratory Teaching*; Goldman, C.A., Ed.; Department of Biology Emory University: Atlanta, GA, USA, 1995; Volume 25, pp. 25–32.
83. Burke, R.D.; Angerer, L.M.; Elphick, M.R.; Humphrey, G.W.; Yaguchi, S.; Kiyama, T.; Liang, S.; Mu, X.; Agca, C.; Klein, W.H.; et al. A Genomic View of the Sea Urchin Nervous System. *Dev. Biol.* **2006**, *300*, 434–460. [CrossRef]
84. Mashanov, V.; Zueva, O.; Rubilar, T.; Epherra, L.; García-Arrarás, J.E. Echinodermata. In *Structure and Evolution of Invertebrate Nervous Systems*; Schmidt-Rhaesa, A., Harzsch, S., Purschke, G., Eds.; Oxford University Press: Oxford, UK, 2016; pp. 665–688.
85. Cobb, J.L.S. Neurobiology of the Echinodermata. In *Nervous Systems in Invertebrates*; Breidbach, O., Kutsch, W., Eds.; Springer US: Boston, MA, USA, 1987; pp. 483–525.
86. Heinzeller, T.; Welsch, U. The Echinoderm Nervous System and Its Phylogenetic Interpretation. In *Brain Evolution and Cognition*; Roth, G., Wullimann, M., Eds.; Wiley-Spektrum: New York, NY, USA, 2001; pp. 41–75.
87. Hyman, L. *The Invertebrates. IV. Echinodermata. The Coelomate Bilateria*; McGraw-Hill Book Co.: New York, NY, USA, 1955.
88. Heinzeller, T.; Welsch, U. Crinoidea. In *Microscopic Anatomy of Invertebrates*; Harrison, F., Chia, F.-S., Eds.; Wiley-Liss: New York, NY, USA, 1994; Volume 14, pp. 1–148.
89. Mashanov, V.S.; Zueva, O.R.; Garcia-Arraras, J.E. Organization of Glial Cells in the Adult Sea Cucumber Central Nervous System. *Glia* **2010**, *58*, 1581–1593. [CrossRef]
90. Kriegstein, A.; Alvarez-Buylla, A. The Glial Nature of Embryonic and Adult Neural Stem Cells. *Annu. Rev. Neurosci.* **2009**, *32*, 149–184. [CrossRef]
91. Malatesta, P.; Appolloni, I.; Calzolari, F. Radial Glia and Neural Stem Cells. *Cell Tissue Res.* **2008**, *331*, 165–178. [CrossRef] [PubMed]
92. Tanaka, E.M.; Ferretti, P. Considering the Evolution of Regeneration in the Central Nervous System. *Nat. Rev. Neurosci.* **2009**, *10*, 713–723. [CrossRef] [PubMed]
93. Zamora, A.J. The Ependymal and Glial Configuration in the Spinal Cord of Urodeles. *Anat. Embryol.* **1978**, *154*, 67–82. [CrossRef] [PubMed]
94. Binyon, J.; Hasler, B. Electrophysiology of the Starfish Radial Nerve Cord. Comp. *Biochem. Physiol.* **1970**, *32*, 747–753.
95. Millott, N.; Okumura, H. The Electrical Activity of the Radial Nerve in *Diadema antillarum* Philippi and Certain Other Echinoids. *J. Exp. Biol.* **1968**, *48*, 279–287. [CrossRef]
96. Cobb, J. The Nervous Systems of Echinodermata: Recent Results and New Approaches. In *The Nervous Systems of Invertebrates: An Evolutionary and Comparative Approach*; Breidbach, O., Kutsch, W., Eds.; Birkhhäuser Verlag: Basel, Switzerland, 1995; pp. 407–424.
97. Mashanov, V.; Zueva, O.; Heinzeller, T.; Dolmatov, I. Ultrastructure of the Circumoral Nerve Ring and the Radial Nerve Cords in Holothurians (Echinodermata). *Zoomorphology* **2006**, *125*, 27–38. [CrossRef]
98. De Bremaeker, N.; Deheyn, D.; Thorndyke, M.C.; Baguet, F.; Mallefet, J. Localization of S1- and S2-like Immunoreactivity in the Nervous System of the Brittle Star Amphipholis squamata (Delle Chiaje 1828). *Proc. R. Soc. B Biol. Sci.* **1997**, *264*, 667–674. [CrossRef] [PubMed]
99. Cobb, J.L.; Stubbs, T.R. The Giant Neurone System in Ophiuroids. I. The General Morphology of the Radial Nerve Cords and Circumoral Nerve Ring. *Cell Tissue Res.* **1981**, *219*, 197–207. [PubMed]
100. Devlin, C.L. The Pharmacology of Gamma-Aminobutyric Acid and Acetylcholine Receptors at the Echinoderm Neuromuscular Junction. *J. Exp. Biol.* **2001**, *204*, 887–896. [CrossRef]
101. Florey, E.; Cahill, M.A. Cholinergic Motor Control of Sea Urchin Tube Feet: Evidence for Chemical Transmission without Synapses. *J. Exp. Biol.* **1980**, *88*, 281–292. [CrossRef]
102. Ryberg, E. The Localization of Cholinesterases and Non-Specific Esterases in the Echinopluteus. *Zool. Scr.* **1974**, *2*, 163–170. [CrossRef]
103. Newman, S.J.; Thorndyke, M.C. Localisation of Gamma Aminobutyric Acid (GABA)-Like Immunoreactivity in the Echinoderm *Asterias rubens*. *Cell Tissue Res.* **1994**, *278*, 177–185. [CrossRef]
104. Wilkie, I.C.; Barbaglio, A.; Maclaren, W.M.; Carnevali, M.D.C. Physiological and Immunocytochemical Evidence That Glutamatergic Neurotransmission Is Involved in the Activation of Arm Autotomy in the Featherstar *Antedon mediterranea* (Echinodermata: Crinoidea). *J. Exp. Biol.* **2010**, *213*, 2104–2115. [CrossRef] [PubMed]
105. Wilkie, I.C.; Barbaglio, A.; Carnevali, M.D.C. The Elusive Role of L-Glutamate as an Echinoderm Neurotransmitter: Evidence for Its Involvement in the Control of Crinoid Arm Muscles. *Zoology* **2013**, *116*, 1–8. [CrossRef]
106. Candia Carnevali, M.; Bonasoro, F.; Thorndyke, M.C.; Patruno, M. Role of the Nervous System in Echinoderm Regeneration. In *Echinoderms 2000*; Barker, M., Ed.; Balkema: Rotterdam, The Netherlands, 2001; pp. 5–20.
107. Inoue, M.; Tamori, M.; Motokawa, T. Innervation of Holothurian Body Wall Muscle: Inhibitory Effects and Localization of 5-HT. *Zool. Sci.* **2002**, *19*, 1217–1222. [CrossRef]

108. Sugni, M.; Ferreri, F.; Bonasoro, F.; Candia Carnevali, M.; Wilkie, I.C. New Evidence for Serotonergic Control of Regenerative Processes in Crinoids. In *Echinoderms: München*; Heinzeller, T., Nebelsick, J., Eds.; Taylor & Francis Group: London, UK, 2004; pp. 141–146.
109. Cottrell, G.A.; Pentreath, V.W. Localization of Catecholamines in the Nervous System of a Starfish, *Asterias rubens*, and of a Brittlestar, Ophiothrix fragilis. *Comp. Gen. Pharmacol.* **1970**, *1*, 73–81. [CrossRef] [PubMed]
110. Díaz-Balzac, C.A.; Mejías, W.; Jiménez, L.B.; García-Arrarás, J.E. The Catecholaminergic Nerve Plexus of Holothuroidea. *Zoomorphology* **2010**, *129*, 99–109. [CrossRef] [PubMed]
111. Howell, E.; Lancaster, A.; Besh, J.; Richardson, B.; Gomez, E.; Harnew-Spradley, S.; Shelley, C. The Dopamine Receptor Antagonist Haloperidol Disrupts Behavioral Responses of Sea Urchins and Sea Stars. *J. Exp. Biol.* **2023**, *226*, 17. [CrossRef] [PubMed]
112. Hoekstra, L.A.; Moroz, L.L.; Heyland, A. Novel Insights into the Echinoderm Nervous System from Histaminergic and FMRFaminergic-Like Cells in the Sea Cucumber *Leptosynapta clarki*. *PLoS ONE* **2012**, *7*, e44220. [CrossRef]
113. Kotsiuba, E.P.; Kotsiuba, A.E. NADPH-Diaphorase Localization in the Radial Nerve Cords of the Starfish *Patiria pectinifera*. *Tsitologiia* **2004**, *46*, 346–351.
114. Martinez, A.; Riveros-Moreno, V.; Polak, J.M.; Moncada, S.; Sesma, P. Nitric Oxide (NO) Synthase Immunoreactivity in the Starfish Marthasterias glacialis. *Cell Tissue Res.* **1994**, *275*, 599–603. [CrossRef]
115. Elphick, M.R.; Melarange, R. Nitric Oxide Function in an Echinoderm. *Biol. Bull.* **1998**, *194*, 260–266. [CrossRef]
116. Elphick, M.R.; Melarange, R. Neural Control of Muscle Relaxation in Echinoderms. *J. Exp. Biol.* **2001**, *204*, 875–885. [CrossRef]
117. Beer, A.J.; Moss, C.; Thorndyke, M. Development of Serotonin-Like and SALMFamide-Like Immunoreactivity in the Nervous System of the Sea Urchin *Psammechinus miliaris*. *Biol. Bull.* **2001**, *200*, 268–280. [CrossRef]
118. Elphick, M.R.; Achhala, S.; Martynyuk, N. The Evolution and Diversity of SALMFamide Neuropeptides. *PLoS ONE* **2013**, *8*, e59076. [CrossRef]
119. O'Neill, P.L. Structure and Mechanics of Starfish Body Wall. *J. Exp. Biol.* **1989**, *147*, 53–89. [CrossRef]
120. Elphick, M.R.; Kemenes, G.; Staras, K.; O'Shea, M. Behavioral Role for Nitric Oxide in Chemosensory Activation of Feeding in a Mollusc. *J. Neurosci.* **1995**, *15*, 7653–7664. [CrossRef] [PubMed]
121. Kaplan, H.M. Anesthesia in Invertebrates. *Fed. Proc.* **1969**, *28*, 1557–1569.
122. Wilkie, I.C. Nervously Mediated Change in the Mechanical Properties of a Brittlestar Ligament. *Mar. Behav. Physiol.* **1978**, *5*, 289–306. [CrossRef]
123. O'Neill, P.L. Torsion in the Asteroid Ray. *J. Morph* **1990**, *203*, 141–149. [CrossRef] [PubMed]
124. O'Neill, P.L. The Effect of Anaesthesia on Spontaneous Contraction of the Body Wall Musculature in the Asteroid *Coscinasterias calamaria*. *Mar. Behav. Physiol.* **1994**, *24*, 137–150. [CrossRef]
125. Motokawa, T.; Wainwright, S.A. Stiffness of Starfish Arm and Involvement of Catch Connective Tissue in the Stiffness Change. Comp. *Biochem. Physiol.* **1991**, *100A*, 393–397.
126. Wilkie, I.C. Nervously Mediated Change in the Mechanical Properties of the Cirral Ligaments of a Crinoid. *Mar. Behav. Physiol.* **1983**, *9*, 229–248. [CrossRef]
127. Wilkie, I.C. Variable Tensility of the Oral Arm Plate Ligaments of the Brittlestar *Ophiura ophiura* (Echinodermata: Ophiuroidea). *J. Zool.* **1992**, *228*, 5–26. [CrossRef]
128. Byrne, M. The Mechanical Properties of the Autotomy Tissues of the Holothurian *Eupentacta quinquesemita* and the Effects of Certain Physico-Chemical Agents. *J. Exp. Biol.* **1985**, *117*, 69–86. [CrossRef]
129. Byrne, M. Induction of Evisceration in the Holothurian *Eupentacta quinquesemita* and the Evidence for the Existence of an Endogenous Evisceration Factor. *J. Exp. Biol.* **1986**, *120*, 25–40. [CrossRef]
130. Wilkie, I.C.; Emson, R.H.; Mladenov, P.V. Morphological and Mechanical Aspects of Fission in *Ophiocomella ophiactoides* (Echinodermata, Ophiuroidea). *Zoomorphology* **1984**, *104*, 310–332. [CrossRef]
131. Wilkie, I.C.; Griffiths, G.V.R.; Glennie, S.F. Morphological and Physiological Aspects of the Autotomy Plane in the Aboral Integument of *Asterias rubens* L. (Echinodermata). In *Echinoderm Research*; De Ridder, C., Dubois, P., Lahaye, M.-C., Jangoux, M., Eds.; A.A. Balkema: Rotterdam, The Netherlands, 1990; pp. 301–313.
132. MacDonald, A.G.; Wann, K.T. *Physiological Aspects of Anaesthetics and Inert Gases*; Academic Press: London, UK, 1978.
133. Jolly, D.W.; Mawdsley-Thomas, L.E.; Bucke, D. Anaesthesia of Fish. *Vet. Rec.* **1972**, *91*, 424–426. [CrossRef] [PubMed]
134. Fraser, D. Understanding Animal Welfare. *Acta Vet. Scand.* **2008**, *50*, S1. [CrossRef]
135. Fraser, D. *Understanding Animal Welfare: The Science in Its Cultural Context*; Wiley-Blackwell Publishing: Ames, IA, USA, 2008.
136. Fraser, D.; Weary, D.M.; Pajor, E.A.; Milligan, B.N. A Scientific Conception of Animal Welfare That Reflects Ethical Concerns. *Anim. Welf.* **1997**, *6*, 187–205. [CrossRef]
137. Lassen, J.; Sandøe, P.; Forkman, B. Happy Pigs Are Dirty!—Conflicting Perspectives on Animal Welfare. *Livest. Sci.* **2006**, *103*, 221–230. [CrossRef]
138. Mason, G.J.; Mendl, M. Why Is There No Simple Way of Measuring Animal Welfare? *Anim. Welf.* **1993**, *2*, 301–319. [CrossRef]
139. Mellor, D.; Patterson-Kane, E.; Stafford, K.J. *The Sciences of Animal Welfare*; John Wiley & Sons: Hoboken, NJ, USA, 2009.
140. Sandøe, P.; Simonsen, H.B. Assessing Animal Welfare: Where Does Science End and Philosophy Begin? *Anim. Welf.* **1992**, *1*, 257–267. [CrossRef]
141. Hemsworth, P.H.; Mellor, D.J.; Cronin, G.M.; Tilbrook, A.J. Scientific Assessment of Animal Welfare. *N. Z. Vet. J.* **2015**, *63*, 24–30. [CrossRef]

142. Bovenkerk, B.; Verweij, M. Between Individualistic Animal Ethics and Holistic Environmental Ethics Blurring the Boundaries. In *Animal Ethics in the Age of Humans*; Bovenkerk, B., Keulartz, J., Eds.; Springer: Cham, Switzerland, 2016; pp. 369–385.
143. Botreau, R.; Veissier, I.; Butterworth, A.; Bracke, M.B.; Keeling, L.J. Definition of Criteria for Overall Assessment of Animal Welfare. *Anim. Welf.* **2007**, *16*, 225–228. [CrossRef]
144. Mellor, D. Enhancing Animal Welfare by Creating Opportunities for Positive Affective Engagement. *N. Z. Vet. J.* **2015**, *63*, 3–8. [CrossRef] [PubMed]
145. Mellor, D. Positive Animal Welfare States and Encouraging Environment-Focused and Animal-to-Animal Interactive Behaviours. *N. Z. Vet. J.* **2015**, *63*, 9–16. [CrossRef] [PubMed]
146. Lahvis, G. Animal Welfare: Make Animal Models More Meaningful. *Nature* **2017**, *543*, 623. [CrossRef]
147. Percy, J.A. Thermal Adaptation in the Boreo-Arctic Echinoid *Strongylocentrotus droebachiensis* (O. F. Müller, 1776). II. Seasonal Acclimatization and Urchin Activity. *Physiol. Zool.* **1973**, *46*, 129–138. [CrossRef]
148. Himmelman, J.H.; Guderley, H.; Vignault, G.; Drouin, G.; Wells, P.G. Response of the Sea Urchin, Strongylocentrotus droebachiensis, to Reduced Salinities: Importance of Size, Acclimation, and Interpopulation Differences. *Can. J. Zool.* **1984**, *62*, 1015–1021. [CrossRef]

animals

MDPI

Article

Assessing Negative Welfare Measures for Wild Invertebrates: The Case for Octopuses

Michaella P. Andrade [1,*], Charles Morphy D. Santos [1], Mizziara M. M. De Paiva [2], Sylvia L. S. Medeiros [2], C. E. O'Brien [3], Françoise D. Lima [4], Janaina F. Machado [5] and Tatiana S. Leite [6]

1. Graduate Program in Evolution and Diversity, Federal University of ABC, Av. dos Estados, 5001, Bairro Bangu, Santo André 09210-580, Brazil; charlesmorphy@gmail.com
2. Graduate Program in Neurosciences, Brain Institute, Federal University of Rio Grande do Norte, Natal 59078-900, Brazil; mizziaradepaiva@gmail.com (M.M.M.D.P.); sylvia.lsmedeiros@gmail.com (S.L.S.M.)
3. The School for Field Studies Center for Marine Resource Studies, Cockburn Harbour TKCA 1ZZ, Turks and Caicos Islands; cobrien@fieldstudies.org
4. OKEANOS, Institute of Marine Sciences, University of the Azores, 9901862 Horta, Portugal; limad.francoise@gmail.com
5. Regional Program for Development and Environment, Federal University of Rio Grande do Norte, Natal 59078-900, Brazil; machadojanainaf@gmail.com
6. Department of Ecology and Zoology, Federal University of Santa Catarina, Florianópolis 88040-900, Brazil; tati.polvo@gmail.com

* Correspondence: michapereirandrade@gmail.com

Simple Summary: Wild octopuses face constant challenges to their survival, such as sublethal predation and conflicts with conspecifics, and can also be affected by human activity (e.g., interactions with divers or with fishing gear) and natural disturbances (e.g., storms, heat waves). These events can impact octopus welfare, especially when they induce stress, facilitate disease transmission, or result in starvation. Understanding the natural behavior of octopuses in the wild can help determine parameters that can be used to assess octopuses' health and welfare. Here, we use photographic and video records of wild octopuses in a variety of negative contexts, with the goal of identifying any potential metrics of welfare. We compare these observations with published data on octopus welfare in captivity and postulate potential consequences of decreased welfare for wild octopuses. Six measures of negative welfare identified for captive animals occurred in wild octopuses as well. We also identified two new measures of negative welfare unique to wild octopuses. This study identified the first set of criteria that can be used to non-invasively assess octopus welfare in the wild. We encourage further development of non-lethal and minimally invasive techniques to quantify welfare in wild invertebrates.

Abstract: Welfare metrics have been established for octopuses in the laboratory, but not for octopuses living in the wild. Wild octopuses are constantly exposed to potentially harmful situations, and the ability to assess the welfare status of wild octopuses could provide pertinent information about individuals' health and species' resilience to stressors. Here, we used underwater photos and videos to identify injuries and stress-related behaviors in wild *Octopus insularis* in a variety of contexts, including interacting with fishermen, interacting with other octopuses and fish, proximity to predators, in den, foraging, and in senescence. We adapted established metrics of octopus welfare from the laboratory to these wild octopuses. In addition to observing all of the stress measures, we also identified two previously unknown measures associated with decreased welfare: (1) a half white eye flash and (2) a half-and-half blotch body pattern. More than half of the individuals analyzed had arm loss, and almost half of the individuals had skin injuries. We also observed that irregular chromatophore expression and abnormal motor coordination were associated with interactions with fishermen. This is the first study to apply measures of welfare from the laboratory to wild octopuses. Our results may also aid in the identification of welfare measures for other wild invertebrates.

Keywords: body patterns; Cephalopoda; field study; sentience

1. Introduction

Animal welfare can be defined as the quality of life experienced by sentient animals [1]. Because octopuses and other cephalopods are animals with highly developed central nervous systems and advanced cognitive abilities, many researchers argue that they are sentient, experiencing physical, mental, and emotional states [2,3] and avoid contexts in which they experience pain [4].

Shallow-water octopuses are generally semelparous with a single reproductive period at the end of their lives. Females produce hundreds to thousands of eggs, of which only a few individuals survive to adulthood [5,6]. As soft-bodied secondary consumers, octopuses are subject to predation, both lethal and sub-lethal (arm loss), from larger bony fish, elasmobranchs, marine mammals, and seabirds [7]. Wild octopuses must constantly weigh the benefit of foraging against the cost of defense and the threat of predation [8] and interactions with other animals [9,10] and must select suitable shelters and perform regular maintenance on this den [11]. Lethal and sublethal predation, fishing, and agonistic and interspecific interactions are some of the daily challenges that can affect the welfare of octopuses [12–14], resulting in physical damage, stress, and/or pain [4].

Standardized measures to quantify overall welfare can be affected by factors including living conditions, diet, and access to enrichment [15]. Measures based on changes in physiology and behavior are used to assess the welfare of vertebrates, but relatively few welfare measures for invertebrates have been validated thus far [16,17]. Differences in physiology, anatomy, and behavior preclude the use of the same welfare measures for vertebrates and invertebrates [16]. Moreover, most research is limited to animal welfare in laboratories and captivity, with the welfare of wildlife having received much less attention [18–20].

The literature describing the association between body patterns and behavioral repertoires in octopuses is still scarce [21]. Most works focus on analyzing body patterns related to camouflage [22], changes in chromatophores [23], and systematics [24]. There is a gap in understanding how much body patterns and their different components can be related to these animals' welfare and emotional states.

Human activities such as hunting and fishing have a detrimental effect on the welfare of many wild animals, resulting in injuries, stress, and pain [18,25]. Although stress and pain can increase fitness in the short term, over time, they can negatively impact animal welfare and health ([19] *contra*, see [15]). The impact of physical injuries on octopus mortality remains poorly understood. Data regarding the effects of injuries and the expression of abnormal behaviors in octopuses offer crucial information about the challenges they face in the wild [19,26–28].

Octopus fishing is a common practice in many coastal communities [29]. In Brazil, octopuses are commonly caught by using a "bicheiro" (an improvised harpoon) [30] but discarded or used as prey for fishing if the individual is too small to sell (Leite, personal report). In fish, the practice of "catch and release" induces, in addition to physical injuries, the release of stress hormones and changes in behavior [31]. The hormonal and behavioral consequences of "catch-and-release" in octopuses should be explored in the coming years, especially with increasing evidence that these animals experience pain, distress, suffering, and lasting harm [4,32–35].

Our study focuses on wild *Octopus insularis* (Leite and Haimovici 2008), a medium-sized benthic species found in the tropical and subtropical southwestern Atlantic. This species has been studied for the past 25 years in Brazil by Projeto Cephalopoda, which has led to the creation of an extensive database of photographs, videos, and scientific information [36–38]. Here, we evaluate photographic and video evidence of injuries and corresponding behavioral changes in wild *O. insularis*. We then compare these data with reports of injuries and concomitant behaviors observed in captive octopuses to isolate potential welfare measures for wild octopuses. In total, we found eight measures related to negative welfare in wild *O. insularis*, six already identified in captive octopuses and two unique to wild animals. This is the first study to apply welfare measures used for captive octopuses to wild ones. Our results are valuable first steps for future in situ investigations seeking to assess welfare and identify the least harmful means of human interaction with wild octopuses.

2. Materials and Methods

Field observations and recording occurred from 2005 to 2023, generating a database of 674 photos and 411 videos of *O. insularis* in various contexts, as well as of octopus fishermen at work. The fieldwork was carried out in several areas of Brazil: Fernando de Noronha National Park; the São Pedro and São Paulo and Abrolhos Archipelagos; the Atol das Rocas Biological Reserve; Nisia Floresta; Maceió; and Salvador. The regions sampled are tropical with a humid coastal climate and high air temperatures throughout the year, ranging between 20 °C and 28 °C. Observational sessions utilized ad libitum and focal follow methods [39] with SCUBA and snorkel. Photographs donated by other researchers, artisanal fishermen, and marine photographers were also utilized.

From the database, we selected media based on photo and video quality; the proportion of the animal's body that was visible; the presence of visible injuries; and for a larger context, the presence of behaviors that could be related to welfare [40]. We collected the following metadata from each photo or video: location, collection date, photographer, number of photographs/videos per animal, behavioral context, number, location and type of skin injuries, number of missing arms, body pattern, presence of irregular chromatophore expression, and abnormal body position. We collected the following data from both photographs and videos: (1) body patterns; (2) type of injuries; (3) number of injuries; (4) irregular chromatophore expression; and (5) abnormal body position. For all variables, except for the number of injuries, we used a binary system to indicate presence, absence, or inability to visualize. In the case of the number of injuries, we quantified both the total number of injuries and the number of missing arms. When analyzing body patterns, we considered the occurrence (presence/absence) of each body pattern in videos, as a single video could contain multiple body patterns. In contrast, for photos, only one body pattern per photo was recorded given the limitation of capturing multiple patterns in a single image.

In describing body patterns and components, we utilized existing cephalopod nomenclature whenever possible [41] (see Supplemental Material for definitions of body patterns). Injuries, context, and behaviors suggestive of impaired welfare present in the database are defined in Table 1 (results). To identify behavioral context, we recorded environmental characteristics and information about the octopuses from the media sources and data collectors' notes.

Table 1. Potential welfare measures in wild *O. insularis*, including injuries, aberrant behaviors or body positions, irregularities in chromatophore expression, and definitions of each negative welfare measure.

Welfare Measures	Definitions	Contexts
Arm loss	Absence of a part of or entire arm(s) or presence of regenerated arm tissue.	Interacting with fishermen, fish, and other octopuses and in den and foraging
Skin injuries	Fixed, irregular, pale, or reddish spots or scars. Often near the eyes/head, arms (excluding whole or partial arm loss), arm web, suckers, or mantle.	Interacting with fishermen, fish, and other octopuses; proximity to a predator; in den and foraging
Irregular chromatophore expression	Uncoordinated color changes or areas of the body with an unchanging uniform color or pattern.	Interacting with fishermen, in den, and foraging
Abnormal body position in water column	Abnormal arm movement and/or loss of muscle tone (e.g., arms).	Interacting with fishermen and entering a state of senescence

Table 1. *Cont.*

Welfare Measures	Definitions	Contexts
Half-and-half blotch	Display of two different body patterns with a distinct longitudinal line dividing them. One half of the body shows the blotch pattern, consisting of white circles on a brown background. The other half of the body shows mottled or brown patterns. The texture is smooth. This pattern can occur when the animal is still or in motion.	Interactions with fish
Excess dermal mucus *	Presence of excess dermal mucus on the mantle and/or parts of the arms, presumably due to a lack of grooming activity (cleaning maneuver).	Foraging
Excess papillae expression *	Papillae are raised for an extended period (at least 1 min).	Foraging
White eye flash *	The area immediately around the pupil flashes pale, usually for two seconds or more. Paleness may occur on both upper and lower halves of the iris or only the lower half.	Foraging

* Reported in a small number (n < 2) of records.

From the eight welfare measures discussed here, we chose four for statistical analysis as they were the most commonly observed in our database (Table 2). The following factors were used as categorical variables: (1) body pattern, (2) number of injuries, (3) irregular chromatophore expression, and (4) abnormal body position. To analyze the association between the contexts and these variables, we used chi-square tests [42] in RStudio version 4.2.1. [43]. We set an alpha of $p < 0.05$, and Bonferroni adjustments were utilized for multiple comparisons [44].

Table 2. The number and percentage of individual *O. insularis* observed in seven contexts potentially impacting welfare.

Context	% Data	Number of Individuals
Interactions with fishermen	42.11%	24
Foraging	22.81%	13
Interactions with fishes	14.04%	8
In den	12.28%	7
Agonistic interactions with other octopus	3.51%	2
Proximity to predator	3.51%	2
Entering a state of senescence	1.75%	1
Total	100%	57

3. Results

3.1. Welfare Measures and Contexts

We found that the total sample was 57 wild octopuses. Of these, 44 octopuses were recorded in 175 photographs, and 13 octopuses were recorded in 23 videos. Our data came from seven different areas in Brazil: 51% from Fernando de Noronha National Park (n = 29), 19% from Nisia Floresta (Rio Grande do Norte) (n = 11), 16% from Atol das Rocas Biological Reserve (n = 9), 7% from Bahia (n = 4), 4% from Macéio (Alagoas) (n = 2), 2% from Abrolhos Archipelago (n = 1), and 2% from São Pedro and São Paulo Archipelago (n = 1).

Eight indicators of welfare were identified in these photos and videos (Table 1). In addition, seven different contexts that could potentially impact welfare were identified: (1) interacting with fishermen (in the act of fishing) (Figure 1); (2) agonistic interactions with other octopuses, either at a distance or via physical contact; (3) interactions with fish; (4) proximity to a predator (when a predator was close by or physically interacting with the octopus); (5) in den (when the octopus was inside a sheltered crevice); (6) foraging (when the octopus was moving along the substrate or swimming in the water column looking for food); and (7) entering a state of senescence (near the end of its lifespan). The highest percentage of observations were from interactions with fishermen (42.11%; n = 24), foraging (22.81%; n = 13), and interactions with fish (14.04%, n = 8) (Table 2).

Figure 1. *Octopus insularis* fishery in Brazil (Bahia). (**A**) An octopus collected by a fisherman; (**B**) an octopus with a skin injury (red arrow) possibly due to a prior capture; (**C**) octopuses being stored after collection by fishermen; and (**D**) an octopus collected with a "bicheiro". Photos by Lucas Ribeiro.

The most commonly utilized potential welfare measure was arm loss (Table 3, Figure 2): 38 individuals (67% of the total sample) were missing at least one arm. Skin injuries were also common, with 24 individuals (42% of the total sample) possessing them. Injuries occurred on a variety of body parts, including the arms, mantle, head, eyes, suckers, and arm web, with some individuals possessing injuries in multiple locations. Irregular chromatophore expression was observed in 15 individuals, 12 of them during interactions with fishermen. There was a significant relationship between injury location and context (χ^2 = 68.64; gl = 42; *p*-value= 0.007). Specifically, during agonistic interactions with conspecifics, more injuries were observed on the mantle, while for individuals in den, more injuries were observed on the head and eyes (z-value = −3.32). Irregular chromatophore expression also varied by context (χ^2 = 48.86; df = 12; *p*-value < 0.0001), more often associated with interactions with fishermen (z-value = −3.04). Abnormal body positions also varied by context (χ^2 = 49.7; df = 12; *p*-value= 0.000001577), most associated with interactions with fishermen (z-value = −3.04).

We identified 82 occurrences within our observations of 57 octopuses in which body patterns could be discerned (Table 4). Approximately half of the patterns were blotched and mottled. Other body patterns observed were uniform dark, uniform light, different variations of longitudinal stripes, half-and-half blotch, and deimatic displays.

Table 3. Number of times each potential welfare indicator was utilized according to context.

	Number of Instances						
Welfare Measures	**Interactions with Fishermen**	**Foraging**	**Interactions with Fish**	**In Den**	**Interactions with Other Octopuses**	**Proximity to Predator**	**Entering a State of Senescence**
Arm loss	21	11	2	2	2	-	-
Skin injuries	13	3	1	4	2	1	-
Irregular chromatophore expression	12	2	0	1	0	0	-
Abnormal body position	9	0	0	0	0	0	1
Total occurrences	55	16	16	7	4	1	1

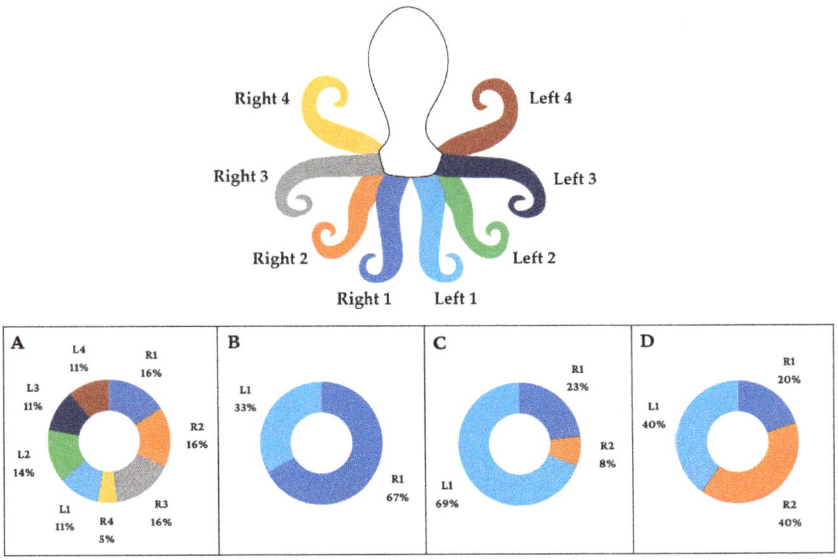

Figure 2. Illustration showing the percentage with which each arm is lost (R1, R2, R3, R4, L1, L2, L3, and L4) and percentage of losses in different contexts: (**A**) interactions with fishermen; (**B**) interactions with fish; (**C**) foraging; and (**D**) interactions with other octopuses. Some individuals had more than one missing arm. The contexts of "in den", "entering a state of senescence", and "proximity to predator" were not included since arm loss was rarely observed in these contexts.

Table 4. Summary of body patterns and components observed in our database ("-" indicates absence).

Body Patterns and Components	**Interactions with Fishermen**	**Foraging**	**Interactions with Fish**	**In Den**	**Interactions with Other Octopuses**	**Proximity to Predator**	**Entering a State of Senescence**
Blotch	16	2	1	-	1	-	-
Mottle	1	10	7	5	-	1	-
Uniform dark	6	3	1	1	1	1	-
Uniform light	3	2	-	1	-	-	1
Light brown	3	1	-	1	-	-	-
Deimatic display	-	1	-	-	-	-	-
Tricolor longitudinal stripes	1	-	-	-	1	-	-
White longitudinal stripe	-	-	1	-	1	-	-
Half and half	2	1	-	-	-	-	-
Half-and-half blotch	-	-	5	-	-	-	-
Total occurrences	32	20	15	8	4	2	1

The body patterns observed in our samples (Table 4) were significantly affected by each of the contexts considered here (χ^2 = 101.77; gl = 54; *p*-value < 0.0001). Our data demonstrated a significant positive relationship between the blotch body pattern and interactions with fishermen, the half-and-half blotch pattern and interactions with fish, and the uniform light pattern and senescence (z-value= −3.38) (Figure 3).

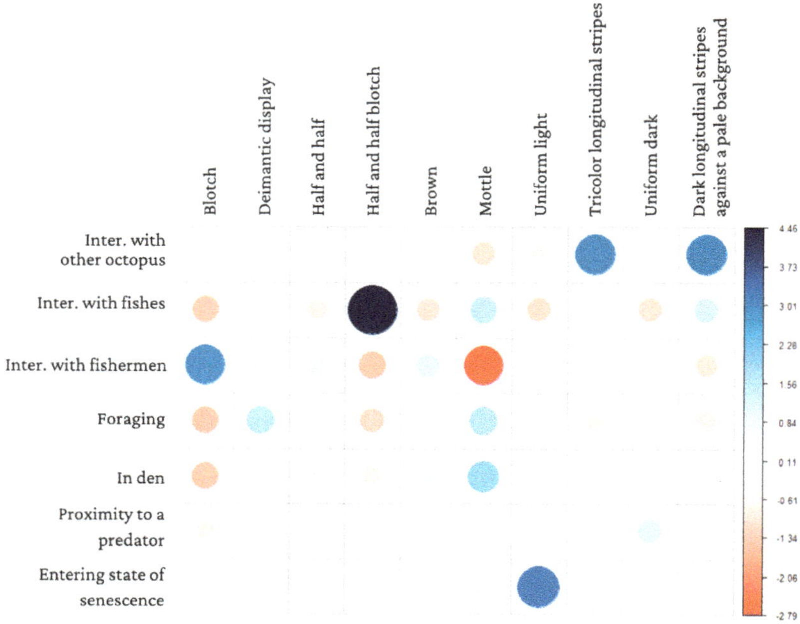

Figure 3. Correlation plot for chi-squared test residuals for each context against body pattern/component (χ^2 = 101.77; gl = 54; *p*-value < 0.0001, 0.00009164). Positive associations are in blue, and negative associations are in red.

3.2. Notes on Octopus Behavior

We noted some interesting behaviors in a few of the individuals we observed in the field. At Boldró Beach (Fernando de Noronha National Park), an octopus was observed with prolonged papillae expression, excessive dermal mucus, and fixed irregular spots of dark chromatophores on the right side of the mantle (Figure 4A–C,E). Right arm I, left arm I, and right arm III were lost, with differing amounts of regenerating tissue. The octopus exhibited a half-and-half body pattern, with the dark side corresponding with the wounded side of the mantle (Figure 4C). We also observed episodes of "white flash" around its pupils, which was sometimes only expressed on the lower part of the above pupil (half white eye flash).

We also recorded a senescent female at Cacimba do Padre Beach (Fernando de Noronha National Park) with pale skin (uniform light) and a near-total loss of muscle tone, as demonstrated by disordered movement of the arms. This female had likely gone through a long period of fasting during the maturation of her eggs (Figure 4T). This is the first record of senescence in a wild female *O. insularis* to date.

At the Atol das Rocas Biological Reserve, we video-recorded an agonistic interaction between two males. The posture of one of the combatants was similar to the "Nosferatu posture" previously reported for *Octopus tetricus* Gould 1852 [45], in which the octopus displays a uniform dark body pattern, with extended arm membranes and a mantle elevated above the rest of the body, moving via backward jetting. We were able to observe the octopuses engaged in "full attack" mode, in which one individual wrapped its arms around the other, and both were thrown against the substrate.

Figure 4. Octopuses under both natural and anthropogenic stressors. (**A**) Dark brown lesion on the right side of the mantle (arrow) indicating the site of a previous injury; (**B**) half white eye flash, pale eye flash on the lower part of the pupil (arrow); (**C**) half-and-half pattern; (**D**) octopus with one arm missing, scratches on the skin (arrows) and expressing a blotch body pattern; (**E**) octopus with uniform dark pattern, loss of arms, and excess dermal mucus (arrows); (**F**) octopus after being caught with a "bicheiro", showing irregular chromatophore expression (arrows); (**G**) tricolor longitudinal

stripes; (**H**) octopus with mottle pattern and wound on one arm (arrow); (**I**) octopus with mottle pattern, raised papillae, and wound on one arm (arrow); (**J**) octopus with an injury on its arms and mantle (arrow); (**K**) an injured octopus lying on a stingray (*Dasyatis americana* Hildebrand and Schroeder, 1928); (**L**) gisherman catching octopus with visible irregular chromatophore expression (arrow); (**M**) octopus after being hit with the fisherman's "bicheiro"; (**N**) lesion above the eye (arrow), white above the pupil, and mottle pattern; (**O**) injury that resulted in the loss of suckers (arrow); (**P**) startle response at an approaching territorial fish, half-and-half blotch body pattern (2/4 s); (**Q**) octopus expressing a deimatic pattern; (**R**) octopus foraging over algae with one missing arm (arrow); (**S**) octopus in den with irregular chromatophore expression: a small portion of the body is expressing the mottle pattern, while the greater portion expresses a pale pattern—normally, these areas would express the same body pattern; and (**T**) senescent female. Photos by S. Medeiros (**A–C**); Buia (**D**); M. Paiva (**E**); L. Ribeiro (**F**); Drauzio (**G**,**P**); T. Leite (**H**,**I**,**R**); Cosme Johnny (**J**); A.T. Souza (**K**); Nego Noronha (**L**,**M**,**O**,**Q**); F. Lima (**N**); H. Bouth (**S**); and Gibson Lemos (**T**).

The "startle response" behavior was seen in eight octopuses in response to territorial fish (mostly in Fernando de Noronha National Park, Nisia Floresta, and Atol das Rocas Biological Reserve). When octopuses were approached by fish, either making physical contact or by rapidly accelerating toward the octopus, they exhibited a half-and-half blotch body pattern, with well-spots on the mantle directed toward the fish (Figure 4P).

Overall, we identified two previously undescribed behaviors in wild octopuses: half-and-half blotch (a startle response) and white eye flash in the lower part of the pupil (derivation of the "eye flash" described in [9]). We also observed the half-and-half body pattern, previously reported only in sleep phases [46], which was expressed when octopuses were awake.

3.3. Comparisons of Welfare Measures between Captive and Wild Octopuses

Based on a review of the literature covering cephalopod welfare in captivity, we compiled a set of potential measures for gauging welfare in wild octopuses (Table 5).

Table 5. Injuries and behaviors described for different octopus species in the laboratory and applicable to wild *Octopus insularis*, along with their putative consequences.

Welfare Measure	Species	Consequences for Wild Octopuses	References
Skin injuries	General for Cephalopoda	Problems with camouflage; increased detection by predators; viral or bacterial diseases; inflammation and hyperalgesia; acute or chronic stress; death.	[47–49]
	Octopus vulgaris, *O. joubini*, *O. briareus*		
Arm loss	General for Cephalopoda	Problems in foraging, locomotion, defense, male competition, and mating success; difficulties in manipulating objects and prey; physical costs beyond energy demands for arm regeneration; death.	[13,47]
	O. bimaculatus, *O. bimaculoides*, *O. rubescens*, *Abdopus* sp.		
Chromatophore failure	General for Cephalopoda	Problems with camouflage; increased predator detection; death.	[47]
Excess dermal mucus	General for Cephalopoda	Parasite accumulation; chronic stress; death.	[47,48]
Extended papillae expression	*O. vulgaris*,	Acute or chronic stress	[40,50]
	Enteroctopus dofleini		
Abnormal body position	*O. bimaculoides*	Chronic stress; problems in foraging, locomotion, and defense; difficulties in manipulating objects and prey; parasite accumulation; death.	[47,51]

We also observed two possible measures not reported before in captive octopuses: (1) the half-and-half blotch body pattern, which seems to indicate that the octopus is experiencing acute stress, and (2) the half white eye flash, associated with excitement and acute stress.

4. Discussion

This is the first study to apply welfare metrics used in captivity to wild octopuses. We identified these metrics using photographs and videos of wild octopuses, an approach already used for other species such as horses [19], whales [52], and penguins [53]. Our study reiterates the efficacy of minimally invasive assessments in the study and pursuit of animal welfare.

Octopus insularis, a diurnal species [54], predominantly inhabits shallow, well-lit waters characterized by high visibility, often exceeding 10 m [6]. These conditions facilitate behavioral studies by allowing individuals to be tracked closely and continuously using cameras placed in front of their dens or in focal follows during excursions from the den [54]. Most of the media we collected were recorded by or during interactions with fishermen, reflecting the fact that *O. insularis* is a highly fished species in Brazil [14,29] and that these animals regularly interact with fishermen in many Brazilian coastal areas. Nevertheless, to date, no study has attempted to understand the impacts of capture and handling by fishermen on free-living octopuses.

We show that all six metrics of negative welfare measures identified for captive octopuses [13,40,47–51] also occur in wild ones. In particular, injuries to the body are quite common. In captive settings, mantle lesions in *Eledone cirrhosa* Lamarck, 1798 have been largely attributed to octopuses "jetting" against tank walls [55]. In the wild, interactions with fishermen and predators and agonistic interactions with conspecifics have the potential to cause injuries. In our sampling, bodily injuries were seen most often in interactions with fishermen and when the octopus was in its den. This may be a protective behavior in response to discomfort, pain [4], or increased vulnerability. Another possibility is that animals with this type of injury need to sleep longer to conserve energy and aid healing, like vertebrates [56].

Arm loss is common in octopuses [13,26,27], especially through predation. Reports have shown a tendency to lose the first left arm (LI) in *O. bimaculatus* Verrill, 1883; *O. bimaculoides* Pickford and McConnaughey, 1949; and *O. rubescens* Berry 1953 [13]. As octopuses use their arms to manipulate prey and objects and to obtain sensory information from the environment [3], arm loss impacts everyday behaviors. For example, arm loss in wild *Abdopus sp.* Norman and Finn 2001 impacts male competition and mating success [57]. Arm loss has been used for welfare assessment in captive octopuses: the loss of three or more arms is the most worrying situation, requiring immediate action (sometimes euthanasia) [40]. In our data, the loss of only one arm was much more frequent than the loss of two, but about a quarter of the animals had lost three or more arms. We observed a tendency to lose arms LI, R1, and R2, suggesting that these arms were used most often in risky behaviors [13,58] or for defense against attacks. Arm loss can have negative implications for the welfare and survival of octopuses in the wild, increasing challenges in foraging and handling food, as well as leading to potential disadvantages in competing with other individuals.

Irregular chromatophore expression was observed in 50% of the interactions between *O. insularis* and fishermen. In addition to skin injuries, the rough handling of these animals can damage their chromatophores. Our field observations strongly suggest that a blow from a "bicheiro" hook can cause chromatophore failure, resulting in areas that are permanently dark or light. Defective chromatophore animals in the wild are probably at a higher risk of predation due to disruption to their camouflage [59].

One of the octopuses displayed a buildup of mucus on its skin, potentially indicative of bacterial growth [27,48]. Normally, an octopus removes excess dermal mucus via grooming, and in laboratory settings, octopuses that cease or reduce grooming are those that show other indicators of poor health [47]. Although our sampling was restricted to a single individual, excessive dermal mucus in wild octopuses has potential consequences for welfare, including an increased likelihood of parasite accumulation on the skin, which can result in chronic stress and even death.

In a captive environment, the excessive expression of papillae on the mantle and above the eyes may indicate excited states in octopuses [40,50]. The exposure of papillae in wild octopuses is related to camouflage and is usually expressed for only short periods of time. We observed an octopus that expressed its papillae for more than 1 min. Although more research is required, this is likely another indicator of poor welfare.

We found a strong association between abnormal motor coordination and interactions with fishermen but not between the octopuses' abnormal motor coordination and foraging. It is worth noting that, during fishing activities, octopuses are often struck by fishing gear, such as "bicheiros", repeatedly without being collected. Octopuses that are particularly degraded may be used as bait for other fishing activities or discarded. The discarded individuals often exhibit anomalous coordination, which is otherwise observed only after egg laying and during senescence [51]. Sometimes, the animal's impaired behavior resembles an elegant "dance" and is recorded and shared on social media [60]. This behavior may be an indication of pain or discomfort. During this abnormal behavior, the muscle tone of the arms visibly decreases, and the animal moves almost exclusively at the whim of the current. This abnormal motor coordination compromises welfare, as it negatively affects the ability to perform basic maintenance behaviors such as grooming, which may lead to the accumulation of parasites on the skin. In the laboratory, abnormal motor behavior is often associated with imminent death [61].

We found two other potential measures of welfare that have not been noted from captive observations: the half-and-half blotch body pattern and the half white eye flash. The blotch components of the half-and-half blotch pattern are characterized by bright white leucophores. Asymmetric body patterns occur in other cephalopods such as cuttlefish during stressful situations like predator approach and during intraspecific competition for mates [62]. The asymmetric behavior that we call half-and-half blotch was expressed in response to territorial fish, which also elicited startle reactions. According to the literature [8,9,62–64], when fish are present around foraging octopuses, this can have both negative and positive effects on their fitness. The negative effects include competition for food [62] and the possibility of alerting predators to their location. However, in at least one instance, the interaction is known to be collaborative [65]. Here, during the half-and-half blotch body pattern, the octopus displayed its blotch pattern toward the approaching territorial fish and the brown or mottle pattern on the other side of the body. A similar situation has also been reported in cuttlefish in potentially stressful situations, such as predator approach and interspecific interactions [62], and in reef squid, where unilateral components appear when fish approach [66]. This is the first description of a unilateral pattern in response to an abrupt approach or touch by a fish in an octopus. Further studies are necessary to determine if the half-and-half blotch body pattern is an acute stress response in all octopuses.

A white flash around the eyes has been previously reported as a response to approaching territorial fish and other disturbances when an octopus is in its den [67], potentially as an anti-predator defense, confusing the predator during a chase [68]. At Boldró Beach, Fernando de Noronha, we recorded a juvenile octopus away from its den (as indicated by a lack of prey remains nearby) that displayed a white flash only on the lower half of its iris (Figure 4B) in response to an approaching snorkeler. This octopus was missing its arms and had an injury on its mantle. It is possible that the white flash is a measure of acute stress in *O. insularis* or else that some injury to his body rendered him unable to shine the upper part of his white eye, but more evidence is needed to confirm this hypothesis.

The investigation of chromatic patterns as potential indicators of stress has been conducted in several captive cephalopods [69–71]. For instance, lighter chromatic patterns, raised papillae, and vertical stripes on the body have been found to be associated with stress in *O. bimaculoides* [65,66]. In this study, we found the uniform light body pattern (seen in the senescent female), the blotch pattern (seen during interactions with fishermen), and the half-and-half blotch pattern (in startle responses in interactions with fishes) to be associated with stress and thus potential visual indicators of stress in *O. insularis*.

Our dataset has limitations primarily arising from the challenges of studying wild octopuses non-invasively. These limitations include difficulties in capturing photos and videos, the absence of unambiguous information about the sex and size of the octopuses, and potential biases introduced by the presence of human observers during recordings.

5. Conclusions

We identified several potential measures of negative welfare in wild *O. insularis* of Brazil, which will enable researchers to visually assess animal conditions in the field. Our results are the first steps for future in situ investigations that require assessments of the welfare of wild octopuses.

The impact of various stressors on the reproductive success and population trends of octopuses remains unstudied. Detailed and systematic knowledge of the factors that affect welfare status in wild octopuses is not only essential for understanding how these potentially sentient animals experience the world but also allows us to predict how different anthropogenic stressors in the marine environment (e.g., pollution, negative interactions with tourism, and fishing) will impact them.

Supplementary Materials: The following supporting information can be downloaded at https://www.mdpi.com/article/10.3390/ani13193021/s1: Table S1: Body pattern and component definitions; Table S2: Table summarizing arm loss table in the octopuses we observed; Table S3: Residuals table of chi-square tests of body patterns by context; Table S4: Residuals table of chi-square tests of injuries by context; Table S5: Residuals table of chi-square tests of irregular chromatophore expressions by context; Table S6: Residuals table of chi-square tests of abnormal body positions by context; Figure S1: Correlation matrix between the variables based on chi-square tests. (A) Body pattern, (B) injuries, (C) irregular chromatophore expression, and (D) abnormal motor coordination. Positive correlations are in blue; negative correlations are in red.

Author Contributions: Conceptualization, M.P.A., C.M.D.S., M.M.M.D.P., S.L.S.M., C.E.O., F.D.L., J.F.M. and T.S.L.; methodology, M.P.A., C.M.D.S., M.M.M.D.P., S.L.S.M., C.E.O., F.D.L., J.F.M. and T.S.L.; investigation, M.P.A., C.M.D.S. and T.S.L.; writing—original draft preparation, M.P.A., C.M.D.S. and T.S.L.; writing—review and editing, M.P.A., C.M.D.S., M.M.M.D.P., S.L.S.M., C.E.O., F.D.L., J.F.M. and T.S.L.; visualization, M.P.A., C.M.D.S., M.M.M.D.P., S.L.S.M., C.E.O., F.D.L., J.F.M. and T.S.L.; supervision, C.M.D.S. and T.S.L.; project administration, M.P.A.; funding acquisition, M.P.A., C.M.D.S. and T.S.L. All authors have read and agreed to the published version of the manuscript.

Funding: This research was conducted as part of the "Investigating sentience, emotional states and personality in wild octopuses" project financed by the Wild Animal Initiative (Grant Number W-8BEN) and Coordenação de Aperfeicoamento de Pessoal de Nível Superior, Brazil (CAPES)—Finance Code 001 (M.P.A.). Funding was also provided by Conselho Nacional de Desenvolvimento Científico e Tecnológico (CNPq), Grant Number: 304027/2022-7 (C.M.D.S.); Fundação Boticário de Proteção a Natureza (0577_20031); CAPES Ciência do Mar II (23038.004807/2014-01); CNPq/SEAP (559863/2008-0); CNPq Pro-Arquipelago (558467/2005-9 and 557157/2009-9); PNPD/Capes (23038.041677/2008-31); PDJ CNPq (155153/2006-4); CNPq Universal (481491-2013-9); Pro-Trindade CNPq (551754/2009-0) (T.S.L.).

Institutional Review Board Statement: Animal welfare regulation in Brazil it is not yet including invertebrates such as cephalopods, and this study did not require ethics committee approval in advance.

Informed Consent Statement: Not applicable.

Data Availability Statement: Photos and videos are unavailable due to privacy restrictions. They are part of the Projeto Cephalopda scientific database.

Acknowledgments: We thank all the collaborators who sent photographs and videos of the wild octopuses and allowed for the elaboration of this work: Lucas Ribeiro, Drauzio, Françoise, Tatiana, Sylvia, Mizziara, Janaina, Buia, Athila, Nego Noronha, Gibson, Helena, Cadu, Cosme Johnny, and A.T. Souza. Thanks to the Wild Animal Initiative; WildStar Films; the staff at ICMBio Noronha, Sea Paradise (Noronha), Maurizélia Brito (Atol das Rocas Biological Reserve), the Cephalopoda Project; and the Other Minds Project team. We also thank Renato Dantas for reviewing the manuscript and providing suggestions, Jennifer Mather for contributing to reflections on the behavior and welfare of wild octopuses, and the reviewers for their relevant considerations in improving this work. C.M.D.S. and T.S.L. would like to thank CAPES and CNPq for grants and scholarships. Permission to study *O. insularis* was granted by ICMBio (License Numbers: 81336-1 and 81336-3, M.P.A.; 10706-5 and 30484-1, F.D.L.; 10706-5, 10706-6, 10706-7, 10706-8, 10706-9, 33754-1, 35405-1, 35405-2, 10706-4, 20109-1, 30484-1, 30484-2, and 30484-2, T.S.L.).

Conflicts of Interest: The authors declare no conflict of interest.

References

1. Mellor, D.J.; Beausoleil, N.J. Extending the 'Five Domains' model for animal welfare assessment to incorporate positive welfare states. *Anim. Welf.* **2015**, *24*, 241–253. [CrossRef]
2. Birch, J.; Burn, C.; Schnell, A.; Browning, H.; Crump, A. Review of the Evidence of Sentience in Cephalopod Molluscs and Decapod Crustaceans. Animal Feeling. 2021. Available online: https://www.lse.ac.uk/News/News-Assets/PDFs/2021/Sentience-in-Cephalopod-Molluscs-and-Decapod-Crustaceans-Final-Report-November-2021.pdf (accessed on 25 July 2023).
3. Mather, J. The Case for Octopus Consciousness: Valence. *NeuroSci* **2022**, *3*, 656–666. [CrossRef]
4. Crook, R.J. Behavioral and neurophysiological evidence suggests affective pain experience in octopus. *iScience* **2021**, *24*, 102229. [CrossRef] [PubMed]
5. Hanlon, R.T.; Messenger, J.B. *Cephalopod Behaviour*; Cambridge University Press: Cambridge, UK, 1996.
6. Leite, T.S.; Haimovici, M.; Mather, J. *Octopus insularis* (Octopodidae), evidences of a specialized predator and a time-minimizing hunter. *Mar. Biol.* **2009**, *156*, 2355–2367. [CrossRef]
7. Dantas, R.J.D.S.; Leite, T.S.; de Albuquerque, C.Q. The trophic interactions of *Octopus insularis* in the food web of a pristine tropical atoll: A baseline for management and monitoring under environmental changes. *Aquat. Ecol.* **2021**, *56*, 269–284. [CrossRef]
8. Meisel, D.V.; Kuba, M.; Byrne, R.A.; Mather, J. The effect of predatory presence on the temporal organization of activity in *Octopus vulgaris*. *J. Exp. Mar. Biol. Ecol.* **2013**, *447*, 75–79. [CrossRef]
9. Mather, J.A. Interactions of juvenile *Octopus vulgaris* with scavenging and territorial fishes. *Mar. Freshw. Behav. Physiol.* **1992**, *19*, 175–182. [CrossRef]
10. Felinto, A.; Mota, E.L.S.; de Souza Rosa, R. Observations of the reef fish *Alphestes afer* (Actinopterygii, Epinephelidae) following the *Octopus insularis* (Cephalopoda, Octopodidae) in a tropical reef system. *Oecologia Aust.* **2020**, *24*, 173–177. [CrossRef]
11. Mather, J.A. 'Home' choice and modification by juvenile *Octopus vulgaris* (Mollusca: Cephalopoda): Specialized intelligence and tool use? *J. Zool.* **1994**, *233*, 359–368. [CrossRef]
12. Mather, J.A.; O'Dor, R.K. Foraging strategies and predation risk shape the natural history of juvenile *Octopus vulgaris*. *Bull. Mar. Sci.* **1991**, *49*, 256–269.
13. Voss, K.M.; Mehta, R.S. Asymmetry in the frequency and proportion of arm truncation in three sympatric California Octopus species. *Zoology* **2021**, *147*, 125940. [CrossRef] [PubMed]
14. Lopes, P.F.; Andrade, L.C.; Pennino, M.G.; Leite, T.S. The inter-annual fishing variability in *Octopus insularis* (Leite & Haimovici 2008) as a result of oceanographic factors. *Fish. Oceanogr.* **2021**, *30*, 515–526. [CrossRef]
15. Browning, H. Welfare comparisons within and across species. *Philos. Stud.* **2023**, *180*, 529–551. [CrossRef]
16. Horvath, K.; Angeletti, D.; Nascetti, G.; Carere, C. Invertebrate welfare: An overlooked issue. *Ann. Dell'istituto Super. Sanità* **2013**, *49*, 9–17. [CrossRef]
17. Carere, C.; Wood, J.B.; Mather, J. Species differences in captivity: Where are the invertebrates? *Trends Ecol. Evol.* **2011**, *26*, 211. [CrossRef] [PubMed]
18. Berg, C.; Lerner, H.; Butterworth, A.; Walzer, C. Wildlife Welfare. *Front. Vet. Sci.* **2020**, *7*, 576095. [CrossRef] [PubMed]
19. Harvey, A.M.; Beausoleil, N.J.; Ramp, D.; Mellor, D.J. A ten-stage protocol for assessing the welfare of individual non-captive wild animals: Free-roaming horses (*Equus ferus caballus*) as an example. *Animals* **2020**, *10*, 148. [CrossRef] [PubMed]
20. Veit, W.; Browning, H. Extending animal welfare science to include wild animals. *Anim. Sentience* **2021**, *1*, 20. [CrossRef]
21. Packard, A.; Hochberg, F.G. Skin patterning in Octopus and other genera. *Symp. Zool. Soc. Lond.* **1977**, *38*, 191–231.
22. Guidetti, G.; Levy, G.; Matzeu, G.; Finkelstein, J.M.; Levin, M.; Omenetto, F.G. Unmixing octopus camouflage by multispectral mapping of *Octopus bimaculoides'* chromatic elements. *Nanophotonics* **2021**, *10*, 2441–2450. [CrossRef]
23. Messenger, J.B. Cephalopod chromatophores: Neurobiology and natural history. *Biol. Rev.* **2001**, *76*, 473–528. [CrossRef] [PubMed]
24. Huffard, C.L. Ethogram of *Abdopus aculeatus* (d'Orbigny, 1834) (Cephalopoda: Octopodidae): Can behavioural characters inform octopodid taxomony and systematics? *J. Molluscan Stud.* **2007**, *73*, 185–193. [CrossRef]

25. Stephen, C.; Wade, J. Wildlife population welfare as coherence between adapted capacities and environmental realities: A case study of threatened lamprey on Vancouver Island. *Front. Vet. Sci.* **2018**, *5*, 227. [CrossRef] [PubMed]
26. Voight, J.R. Movement, injuries and growth of members of a natural population of the *Pacific pygmy* octopus, *Octopus digueti*. *J. Zool.* **1992**, *228*, 247–264. [CrossRef]
27. Gestal, C.; Pascual, S.; Guerra, Á.; Fiorito, G.; Vieites, J.M. *Handbook of Pathogens and Diseases in Cephalopods*; Springer: Berlin/Heidelberg, Germany, 2019. [CrossRef]
28. Mellor, D.J.; Beausoleil, N.J.; Littlewood, K.E.; McLean, A.N.; McGreevy, P.D.; Jones, B.; Wilkins, C. The 2020 five domains model: Including human–animal interactions in assessments of animal welfare. *Animals* **2020**, *10*, 1870. [CrossRef]
29. Leite, T.S.; Haimovici, M.; Oliveira, J.E.L. Uma proposta de manejo para a pesca do polvo *Octopus insularis* Leite & Haimovici, 2008 (Mollusca: Cephalopoda) no Arquipélago de Fernando de Noronha, Brasil. *Arq. Ciên. Mar.* **2008**, *41*, 81–89.
30. Sauer, W.H.; Gleadall, I.G.; Downey-Breedt, N.; Doubleday, Z.; Gillespie, G.; Haimovici, M.; Fecl, G. World octopus fisheries. *Rev. Fish. Sci. Aquac.* **2021**, *29*, 279–429. [CrossRef]
31. Cooke, S.J.; Sneddon, L.U. Animal welfare perspectives on recreational angling. *Appl. Anim. Behav. Sci.* **2007**, *104*, 176–198. [CrossRef]
32. Rowell, C.H.F. Activity of interneurones in arm of octopus in response to tactile stimulation. *J. Exp. Biol.* **1996**, *44*, 589–605. [CrossRef]
33. Hague, T.; Florini, M.; Andrews, P.L.R. Preliminary in vitro functional evidence for reflex responses to noxious stimuli in the arms of *Octopus vulgaris*. *J. Exp. Mar. Biol. Ecol.* **2013**, *447*, 100–105. [CrossRef]
34. Alupay, J.S.; Hadjisolomou, S.P.; Crook, R.J. Arm injury produces long-term behavioral and neural hypersensitivity in octopus. *Neurosci. Lett.* **2014**, *558*, 137–142. [CrossRef] [PubMed]
35. Perez, P.V.; Butler-Struben, H.M.; Crook, R.J. The selective serotonin reuptake inhibitor fluoxetine increases spontaneous afferent firing, but not mechannociceptive sensitization, in octopus. *Invertebr. Neurosci.* **2017**, *17*, 10. [CrossRef] [PubMed]
36. Leite, T.S.; Mather, J.A. A new approach to octopuses' body pattern analysis: A framework for taxonomy and behavioral studies. *Am. Malacol. Bull.* **2008**, *24*, 31–41. [CrossRef]
37. Batista, A.T.; Leite, T.S. *Octopus insularis* (Cephalopoda: Octopodidae) on the tropical coast of Brazil: Where it lives and what it eats. *Braz. J. Oceanogr.* **2016**, *64*, 353–364. [CrossRef]
38. Bouth, H.F.; Leite, T.S.; Lima, F.D.D.; Oliveira, J.E.L. Atol das Rocas: An oasis for *Octopus insularis* juveniles (Cephalopoda: Octopodidae). *Zoologia* **2011**, *28*, 45–52. [CrossRef]
39. Altmann, J. Observational study of behavior: Sampling methods. *Behaviour* **1974**, *49*, 227–266. [CrossRef]
40. Holst, M.M.; Miller-Morgan, T. The Use of a species-specific health and welfare assessment tool for the giant pacific octopus, *Enteroctopus dofleini*. *J. Appl. Anim. Welf. Sci.* **2021**, *24*, 272–291. [CrossRef]
41. Mather, J.A.; Alupay, J.S. An ethogram for Benthic Octopods (Cephalopoda: Octopodidae). *J. Comp. Psychol.* **2016**, *130*, 109. [CrossRef]
42. Sokal, R.R.; Rohlf, F.J. *Biometry: The Principles and Practice of Statistics in Biological Research*. W.H. Freeman and Company: San Francisco, CA, USA, 1981; Volume 2, p. 776. [CrossRef]
43. Team, R. *RStudio: Integrated Development for R.*; RStudio Inc.: Boston, MA, USA, 2015; Volume 700, p. 879.
44. Sokal, R.R.; Rohlf, F.J. *Biometry: The Principles and Practice of Statistics in Biological Research*. W.H. Freeman and Company: San Francisco, CA, USA, 1995; Volume 3, p. 887.
45. Scheel, D.; Godfrey-Smith, P.; Lawrence, M. Signal use by octopuses in agonistic interactions. *Curr. Biol.* **2016**, *26*, 377–382. [CrossRef]
46. Medeiros, S.; de Paiva, M.M.M.; Lopes, P.H.; Blanco, W.; de Lima, F.D.; de Oliveira, J.B.C.; Ribeiro, S. Cyclic alternation of quiet and active sleep states in the octopus. *iScience* **2021**, *24*, 102223. [CrossRef]
47. Fiorito, G.; Affuso, A.; Basil, J.; Cole, A.; de Girolamo, P.; D'angelo, L.; Andrews, P.L. Guidelines for the care and welfare of cephalopods in research–a consensus based on an initiative by CephRes, FELASA and the Boyd Group. *Lab. Anim.* **2015**, *49* (Suppl. S2), 1–90. [CrossRef] [PubMed]
48. Farto, R.; Armada, S.P.; Montes, M.; Guisande, J.A.; Pérez, M.J.; Nieto, T.P. *Vibrio lentus* associated with diseased wild octopus (*Octopus vulgaris*). *J. Invertebr. Pathol.* **2003**, *83*, 149–156. [CrossRef] [PubMed]
49. Hanlon, R.T.; Forsythe, J.W.; Cooper, K.M.; Dinuzzo, A.R.; Folse, D.S.; Kelly, M.T. Fatal penetrating skin ulcers in laboratory-reared octopuses. *J. Invertebr. Pathol.* **1984**, *44*, 67–83. [CrossRef] [PubMed]
50. Andrews, P.L.R.; Messenger, J.B.; Tansey, E.M. The chromatic and motor effects of neurotransmitter injection in intact and brain-lesioned Octopus. *J. Mar. Biol. Assoc. U. K.* **1983**, *63*, 355–370. [CrossRef]
51. Wang, Z.Y.; Ragsdale, C.W. Multiple optic gland signaling pathways implicated in octopus maternal behaviors and death. *J. Exp. Biol.* **2018**, *221*, jeb185751. [CrossRef]
52. Boys, R.M.; Beausoleil, N.J.; Pawley, M.D.; Betty, E.L.; Stockin, K.A. Evaluating potential cetacean welfare indicators from video of live stranded long-finned pilot whales (*Globicephala melas edwardii*). *Animals* **2022**, *12*, 1861. [CrossRef]
53. Freire, R.; Massaro, M.; McDonald, S.; Trathan, P.; Nicol, C.J. A citizen science trial to assess perception of wild penguin welfare. *Front. Vet. Sci.* **2021**, *8*, 698685. [CrossRef]
54. O'Brien, C.E.; Medeiros, S.L.D.S.; Leite, T. Octospy: What *Octopus insularis* do in their dens. *Mar. Ecol.* **2023**, e12763. [CrossRef]

55. Smith, L.E.; Rowe, C.; Mackay, F.; Matthews, C.; Matthews, C.G. Aquarium tank design is integral to the elimination of mantle abrasion in the captive Curled Octopus (*Eledone cirrhosa*): A case study at Macduff Marine Aquarium. *J. Appl. Anim. Welf. Sci.* **2022**, *25*, 355–361. [CrossRef]
56. Smith, T.J.; Wilson, M.; Karl, J.P.; Orr, J.; Smith, C.; Cooper, A.; Montain, S.J. Impact of sleep restriction on local immune response and skin barrier restoration with and without "multinutrient" nutrition intervention. *J. Appl. Physiol.* **2018**, *124*, 190–200. [CrossRef]
57. Wada, T. Size-assortative mating and arm loss in the wild shallow-water octopus *Abdopus* sp. (Cephalopoda: Octopodidae). *J. Nat. Hist.* **2017**, *51*, 2635–2644. [CrossRef]
58. Byrne, R.A.; Kuba, M.J.; Meisel, D.V.; Griebel, U.; Mather, J.A. Does *Octopus vulgaris* have preferred arms? *J. Comp. Psychol.* **2006**, *120*, 198. [CrossRef] [PubMed]
59. Hanlon, R.T.; Forsythe, J.W.; Joneschild, D.E. Crypsis, conspicuousness, mimicry and polyphenism as antipredator defences of foraging octopuses on Indo-Pacific coral reefs, with a method of quantifying crypsis from video tapes. *Biol. J. Linn. Soc.* **1999**, *66*, 1–22. [CrossRef]
60. Instagram. Available online: https://www.instagram.com/p/CrDhrD6p5iR/ (accessed on 3 August 2023).
61. Holst, M.M.; Hauver, C.M.; Stein, R.S.; Milano, B.L.; Levine, L.H.; Zink, A.G.; Watters, J.V.; Crook, R.J. Behavioral changes in senescent giant Pacific octopus (*Enteroctopus dofleini*) are associated with peripheral neural degeneration and loss of epithelial tissue. *Comp. Biochem. Physiol. Part A Mol. Integr. Physiol.* **2022**, *271*, 111263. [CrossRef]
62. Langridge, K.V. Symmetrical crypsis and asymmetrical signalling in the cuttlefish *Sepia officinalis*. *Proc. R. Soc. B Biol. Sci.* **2006**, *273*, 959–967. [CrossRef]
63. Diamant, A.; Shpigel, M. Interspecific feeding associations of groupers (Teleostei: Serranidae) with octopuses and moray eels in the Gulf of Eilat (Agaba). *Environ. Biol. Fishes* **1985**, *13*, 153–159. [CrossRef]
64. Krajewski, J.P.; Bonaldo, R.M.; Sazima, C.; Sazima, I. Octopus mimicking its follower reef fish. *J. Nat. Hist.* **2009**, *43*, 185–190. [CrossRef]
65. Somaweera, R.; Somaweera, R. Nuclear–follower foraging behaviour between Western Australian common octopus and brown-spotted wrasse. *Mar. Freshw. Res.* **2021**, *72*, 1679–1681. [CrossRef]
66. Mather, J.A. Vigilance and antipredator responses of Caribbean reef squid. *Mar. Freshw. Behav. Physiol.* **2010**, *43*, 357–370. [CrossRef]
67. Mather, J.A.; Mather, L. Skin colours and patterns of juvenile *Octopus vulgaris* (Mollusca, Cephalopoda) in Bermuda. *Vie Milieu/Life Environ.* **1994**, *44*, 267–272.
68. Loeffler-Henry, K.; Kang, C.; Yip, Y.; Caro, T.; Sherratt, T.N. Flash behavior increases prey survival. *Behav. Ecol.* **2018**, *29*, 528–533. [CrossRef]
69. Wood, J.B.; Wood, D.A. Enrichment for an advanced invertebrate. *Shape Enrich.* **1999**, *8*, 1–5.
70. Beigel, M.; Boal, J. The effect of habitat enrichment on the mudflat octopus. *Shape Enrich* **2006**, *15*, 3–6.
71. Yasumuro, H.; Ikeda, Y. Effects of environmental enrichment on the behavior of the tropical octopus *Callistoctopus aspilosomatis*. *Mar. Freshw. Behav. Physiol.* **2011**, *44*, 143–157. [CrossRef]

animals

MDPI

Review

It Is Not Only Data—Freshwater Invertebrates Misused in Biological Monitoring

Paweł Koperski

Institute of Functional Biology and Ecology, Faculty of Biology, University of Warsaw, Żwirki i Wigury 101, 02-089 Warszawa, Poland; p.t.koperski@uw.edu.pl

Simple Summary: The article explains why using free-living invertebrates in biomonitoring aimed at assessing the ecological status of rivers, lakes, streams and ponds can be considered misuse. Invertebrates are excluded from ethical considerations in environmental procedures, resulting in the killing of many more individuals than necessary during such activities. Biomonitoring used as a routine method of environmental protection causes cruel deaths of up to millions of aquatic animals every year. Improperly planned procedures which result in excessive mortality have or may have a negative impact on the environment and biodiversity. The life of aquatic invertebrates, although they should be considered sensitive beings, is reduced to an informative function; they become only data useful for biomonitoring purposes. Some new methods, modifications and improvements of biomonitoring procedures that can significantly reduce freshwater invertebrate mortality are presented in the section "Future Directions". Especially the development of effective, precise and reliable methods of survival, e.g., based on the analysis of DNA taken directly from the environment (eDNA), seems to be not only a breakthrough in biomonitoring, but also an important step towards a significant improvement in the welfare of aquatic invertebrates.

Abstract: The article presents and discusses the issues of the use of free-living invertebrates to assess the ecological status of freshwater environments with different methods of biological monitoring. Invertebrates are excluded from ethical consideration in the procedures of environmental protection, which results in the killing of many more individuals during sampling than necessary. Biomonitoring is used as a routine method for environmental protection that results in the cruel death of even millions of aquatic animals annually. In many cases, the mortality of animals used in such types of activities has been shown as excessive, e.g., because the vast majority die due to unnecessary subsampling procedures. Improperly planned and conducted procedures which result in excessive mortality have or may have a negative impact on the environment and biodiversity. Their existence as sensitive beings is reduced to an information function; they become only data useful for biomonitoring purposes. The main problem when trying to determine the mortality of invertebrates due to biomonitoring activities and its impact on natural populations seems to be the lack of access to raw data presenting how many animals were killed during sampling.

Keywords: biological assessment; animal ethics; methods; mortality; nature protection; welfare

1. Introduction

"It is clear that we have direct ethical obligations to sentient animals; but it is not at all clear that we have direct ethical obligations to entities such as species, or to biological diversity. The burden of proof should thus be on conservationists to show how killing the first to preserve the second can possibly be acceptable from an ethical point of view" [1]. Biological monitoring (biomonitoring), in the broadest sense, can be defined as activities aimed at assessing the condition of the environment, performed according to an agreed methodology and using bioindicators. It consists of the assessment of different parameters and detecting ongoing changes in ecosystems and components of biological diversity, including types of natural habitats, populations and species, as well as to assess the effectiveness of the nature protection methods used. In this approach, biological monitoring aims not only to assess the state of the natural environment, organisms and ecosystems, but also other environmental parameters, such as the degree of habitat transformation, and soil, air and water pollution. Some methods are based on the presence or relative abundance of specific taxonomic groups in the environment. The assumption of some of the biomonitoring methods, especially those aimed at assessing the presence of xenobiotics in the environment and their effects, depends on the collection of whole organisms, body fragments or various physiological secretions for analysis.

The Convention on Biological Diversity [2] encourages states that are parties to it to monitor the elements of biological diversity, with particular emphasis on its most endangered components and those representing the greatest potential value for sustainable use. This should include, in particular, monitoring the effects of processes and activities that have or may have a significant negative impact on the conservation and sustainable use of biological diversity. According to the convention, environmental monitoring should cover all levels of biodiversity from ecosystems, through to species level, to genetic diversity.

Since the 1980s, this type of activity has been most intensively carried out in freshwater environments. In European Union countries, in accordance with the requirements of the Water Framework Directive of the European Parliament (WFD) [3], the basis for this type of assessment is the analysis of the composition of the so-called "biological elements". This term means groups of organisms with a confirmed high indicative value. Indicator organisms can be used for biological monitoring purposes at various organizational levels, from the sub-organismal level (e.g., genes, cells, tissues) through the organismal level, to populations, communities or even a whole ecosystem level [4]. Among the five biological elements used to assess the ecological status of European freshwater environments: fish, macrophytes, phytoplankton, periphyton algae and macroinvertebrates, the latter is the most often and most commonly used. Macroinvertebrates remain on the sieve with a mesh size of 0.2–0.5 mm, i.e., in practice have a body length exceeding 1 mm. It is estimated that about two-thirds of the methods for assessing the quality of flowing waters are based on benthic macroinvertebrates.

About 15,000 species of Metazoa occur in European waters, of which about 10,500 can be termed macroinvertebrates. The vast majority of those used in monitoring programs are insects, in particular their larval forms. However, the less numerous groups include water-mites (Acari: Arachnida), snails and mussels (Mollusca), leeches and oligochaete worms (Clitellata), malacostracan and other groups of so-called Crustacea. Insects contain the taxa with the highest indicative value e.g., the EPT group (Ephemeroptera, Plecoptera, Trichoptera larvae). Many species of freshwater invertebrates in Europe are endangered [5] and are particularly sensitive to habitat change as flow alterations, habitat fragmentation, long-lasting drought and pollution are their main threats [6]. The recently intensively studied global decrease in the number and biomass of insects, including ecologically specialized rare species of aquatic insects, is undoubtedly related to the degradation of freshwater environments [7].

A significant part of the waters located in areas of intensive human impact are degraded, and their functional, hydrological and chemical parameters as well as the taxonomic composition of the organisms inhabiting them significantly differ from the pristine conditions. Routine assessment of the ecological status of these environments is carried out at designated sites by official institutions appointed by the authorities to collect data on ecosystem degradation. Freshwater environments—lakes, ponds, wetlands, streams and rivers—are some of the most threatened habitat types on Earth. They contain less than three percent of the volume of water stored on Earth, but about ten percent of animal species live in them. Typically, a large part of terrestrial biodiversity is concentrated near freshwater [8]. Declining numbers of aquatic invertebrates have negative effects on ecosystems because the populations of many species are essential for their function. These organisms filter a huge amount of edible particles suspended in water, control prey populations as predators, constitute the food base of fish species and consume periphytic algae and dead organic matter, which accelerates the biomass turnover. The activity of certain groups reduces the effects of eutrophication and intensifies the self-purification processes in watercourses. Huge swarms of winged mayflies, true-flies and caddisflies after their emergence constitute an irreplaceable food base for many terrestrial vertebrates [9]. As significant reductions in their abundance may cause disturbances in the functioning of the ecosystem, it means that excessive mortality can be regarded as a kind of environmental degradation.

How should we understand the statement that invertebrates are 'misused' in the title of the article and what is this 'misuse' supposed to consist of? In our opinion, this misuse means, above all, excessive suffering and excessive mortality as a result of biomonitoring procedures that affect a large number of aquatic animals. Most often it takes the form of non-selective bulk sampling with the use of sieves, dip-nets, various samplers and toxic substances with a preservative effect. As a result of such activities, the number of animals sampled and killed exceeds, sometimes many times, the numbers necessary to properly apply the ecological status assessment protocols. Researchers often kill far more animals in each sampling effort than necessary to conduct a proper assessment of the ecological status of the habitat for fear of collecting too little data. Reducing suffering and reducing mortality is generally not a primary or secondary objective underlying any official assessment methods. The term 'excessive mortality' should be considered in detail. From an ethics point of view when treating animals as sentient beings (animal welfare), 'excessive' can be generally referred to any number of beings that we endow with inherent moral value, without attempting to prevent it. From the point of view of nature conservation, the meaning of this term is more difficult to define. It certainly concerns a situation where the level of animal mortality permanently threatens the stability of the population. In a situation such as that described in this paper, however, it is impossible to even attempt to assess the strength of such an influence. So, it seems justified to use the term 'excessive mortality' to describe the effects of sampling methods in which many thousands of specimens are killed non-selectively, most of which are not used later for any purpose [10].

Use of Freshwater Invertebrates in Biomonitoring

A typical method of assessing the ecological status of a freshwater environment, in accordance with the requirements of the WFD, is based on the following points: (i) selection of the field site, as assigned to the appropriate abiotic type; (ii) quantitative sampling of invertebrates from the appropriate bottom area, using specialized equipment; (iii) preserving the samples using a suitable substance; (iv) random sampling in the laboratory from the preserved organisms of sub-samples of the total sample until the appropriate number of specimens are obtained, e.g., the minimum for proper evaluation; (v) identification of animals selected in sub-samples to the required level (typically the level of family); (vi) calculation of the value of the multimetric biotic index on the basis of taxonomic composition and richness of the invertebrate fauna; (vii) determination of the class of the ecological status (on a five-point scale) on the basis of the final index score depending on the habitat type [11]. A detailed description of the protocol is available online [12].

It should be noted that new versions of protocols developed for the assessment of freshwater environments increasingly contain propositions intended to reduce the mortality of at least some invertebrates, without decreasing the quality of the assessment. There, you can find suggestions to return to the environment as many individuals as possible whose identification is possible already in the field, e.g., mussels from the Unionidae family, crayfish, as well as to review samples in the field for the presence of protected species and if they occur, their presence should be recorded in field protocol, and the animals removed from the sample and left in the environment [12]. In practice, during field analysis, such a procedure may be only partially effective, e.g., due to the inability to see the younger (smaller) developmental stages of large animals and the inability to distinguish in situ individuals of protected and unprotected species (e.g., dragonflies or beetles).

The methods of biological assessments of freshwater environments differ in both the sampling methods and the biological variables used [13]. Most published studies and reports for different reasons do not mention the total number of animals killed (preserved) but only the number of animals used for analysis. Therefore, it is very difficult to directly compare the levels of invertebrate mortality during routine biological monitoring carried out by different methods. Important factors affecting such mortality rates are: sampling area, methods, intensity, mesh size, sample replications and handling methods of trapped animals (live sorting, preservation of the whole sample, non-lethal collection or identification of selected taxa). Some procedures which are applied in the laboratory post-preservation (i.e., the use of magnification during sorting, sub-sampling and the level of identification) appear to have no effect on mortality, because they only reduce the number of animals used for analysis after preservation, and thus killing all; regardless of the method, it is never an easy "humanitarian" death [11].The effects of using the described sampling methodology in the routine monitoring of flowing waters in Poland between 2011 and 2018 were analyzed in detail [10,11]—this will be used to illustrate general patterns. The methods of biomonitoring in lakes, dam reservoirs and coastal brackish waters used in Poland are so similar that the effects caused by them are probably similar [12]. It can be assumed that the conclusions from a detailed analysis of the effects of sampling in routine monitoring in other countries using methods based on similar assumptions and defined by the requirements of the WFD would be similar, e.g., [14,15]. From the point of view of the considerations discussed in this article, the most important direct effects of the described procedure of preserving entire bulk samples and the use of subsample analyses in the monitoring of flowing waters are:

1. The number of aquatic invertebrates killed (preserved) during sampling is sometimes very high, which significantly exceeds the minimum number necessary for a correct assessment of the state of the environment;
2. A significant number of the animals killed during conservation are not used for analysis, which receives much less abundant sub-samples;
3. There are significant variations in the numbers of animals killed during maintenance, involving the taking of samples for analysis by staff from different laboratories, following the same protocols and often in very similar types of environments.

The number of animals killed during the sampling procedure exceeds, on average, 12 times the number necessary for proper analysis (the described method requires placing at least 350 identified individuals in the database intended for the calculation of the index). In sixty-one percent of the samples, at least five times as many animals were killed than the number sufficient for analysis, in some cases reaching even over 200–500 times more. Moreover, it can be estimated that these numbers are actually even higher due to the fact that for taxa occurring in high densities, the procedure allowed to record only the first 100 individuals sampled, but typically it is not noted in the archived data [12].

As a result of the described subsampling procedure, 80.4% of the animals killed as a result of conservation were not used for the assessment of the environment and died, so to speak. Among the animals of which this type of misuse applied, insects (63.4% of individuals) and Malacostraca (17.3%) predominate. Identifying animals only at the family level means that the collected data cannot be used for scientific purposes (e.g., biodiversity analysis) and it also makes it impossible to reliably assess the impact of the procedure on the populations of rare and protected species. For example, over 10,000 individuals of the mussel family Unionidae were killed, among which there are rare, vanishing and protected species, of which about 8300 individuals, despite being killed, were not included in the analysis. In the case of the Heptageniidae family (Ephemeroptera, mayflies), which also includes rare, endangered and protected species, these numbers are 125,455 (killed specimens), 24,577 (number enough for analysis) and 100,878 (not used for analysis), respectively. In total, there were 2,817,777 specimens which belonged to families, including species covered by various forms of legal protection.

It has been estimated that between 2012 and 2019, at least 12.7 million invertebrates (more than 8.6 million aquatic insects) fell victim to biomonitoring in Poland's watercourses (Figure 1), and according to very rough estimates, on a European scale it was about 18 million animals per year [10,11]. It is very difficult to assess what effects such mortality has on aquatic invertebrate populations and the functioning of ecosystems. This type of impact can only be expected locally in specific environments or directed in isolated populations living in low densities. The differences in the average numbers of invertebrates killed between laboratories performing routine biomonitoring can be extreme (up to 35 times). Importantly, this does not usually stem from the effect of differences in animal densities, resulting from the ecological and geological differences between environments. The average number of invertebrates collected by the staff of different laboratories from one square meter patches at the bottom of watercourses of the same abiotic type (and thus very similar in terms of hydrological, geological and ecological parameters) differed even up to thirty-two times. In the analysis, much attention was paid to the hypothetical reasons for such large differences in mortality values during sampling [10]. The overall variance associated with the total number of insects killed during sampling is explained mainly by: (i) the specificity of the laboratory and (ii) the specificity of the abiotic type; the first factor is much more important than the second one. The remaining analyzed factors are of very little importance. This clearly indicates that it is the "human factor" that is difficult to define—the level of empathy in dealing with invertebrates, which varies greatly among individual people taking samples, is the most important in this case. More important than ecological factors, freedom in the use of the sampling equipment allowed by the procedure and even the aforementioned lack of mechanisms limit the maximum sample abundance.

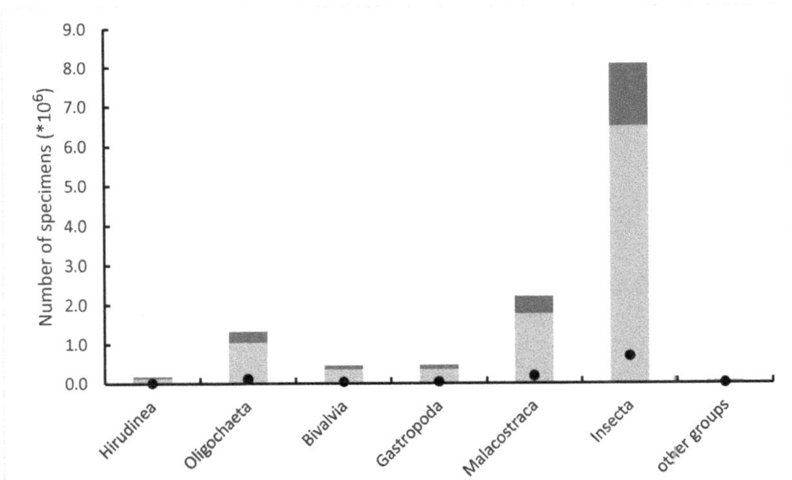

Figure 1. Estimated number of animals from the most important taxonomic groups that were killed during sampling for the purposes of biomonitoring in Polish watercourses in 2012–2019, with the number of specimens used for the analysis (dark grey) and the number of individuals necessary for the correct analysis (black dots) shown.

2. Discussion

Freshwater invertebrates are still treated as non-sentient, thus are devoid of intrinsic value and therefore do not require any form of ethical protection. Deep differences in the endowment of vertebrates and invertebrates with moral entitlement are still clearly visible and take the form of a clear asymmetry [16]. Ethical conflicts are inevitable as a result of large-scale biomonitoring activities resulting in the mass non-selective killing of invertebrates. It seems that the main reason is the simultaneous classification of the animals used into various, often inseparable, functional categories, to which moral rights are granted (or not) to varying degrees. The simplified effects of such classification are shown in Figure 2. Recognizing a given group of animals only as "elements of the population with high indicative value in biomonitoring procedures" automatically excludes them from the sphere of ethical protection and in practice reduces their existence only to an information function—their existence as sensitive beings and all of their life's interests from that point on become just data for the purposes of biomonitoring. In other words, it deprives them of an intrinsic ethical value, giving them only an instrumental value, significant only from the point of view of human benefit. Removing invertebrates from the sphere of ethical protection makes their killing during monitoring an ethically neutral activity, regardless of the number of animals killed. It should be emphasized that such commonly held beliefs about relations between humans and animals are usually based on a personal worldview shaped by religious principles, as argued, e.g., [17,18]. Probably for some people participating in this type of activity, obtaining data by counting the bodies of killed animals does not differ, in ethical terms, from the assessment of the concentration of microorganisms or plant biomass in water, or even the concentration of chemical compounds dissolved in water [11]; it then becomes an act perceived positively.

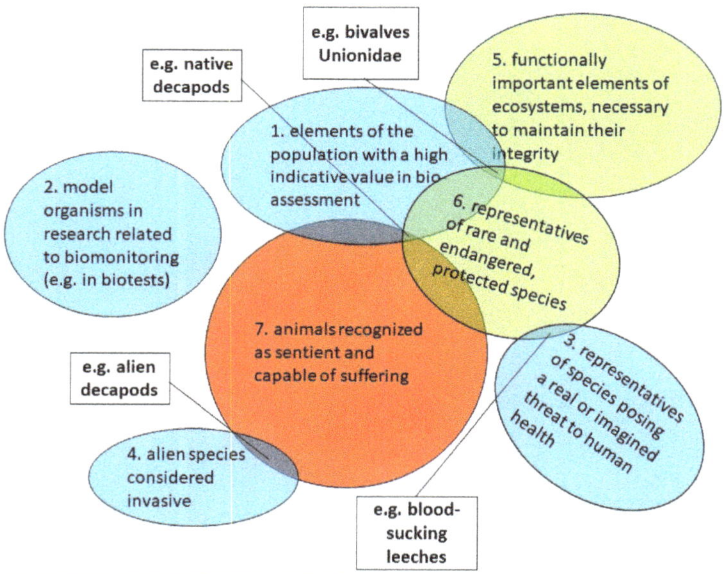

Figure 2. Freshwater invertebrates used in biomonitoring assigned to different functional categories that differ in the moral entitlements granted to them by humans (blue figures—no moral entitlements, green—some moral entitlements granted on the basis of environmental ethics (as exemplars of species), red—moral entitlements granted on the basis of animal ethics (as sentient beings). Ethical conflicts arise where figures of different colors overlap—examples of such animals are shown in empty rectangles.

In the classic twentieth-century publications focused on animal ethics, questions of moral duty are limited to homoiothermic vertebrates. It is necessary to pay attention to the real revolution of ideas that we have been dealing with in the last twenty years, with the results of studies in the fields of neuroscience and behavioral biology clearly demonstrating the presence of advanced cognitive functions, individual behavioral types and emotions in invertebrates. According to them, empirical evidence unequivocally commands abandoning the habit of treating invertebrates as primitive beings whose behavior is based solely on simple reactions to stimuli and automated instincts [19,20], and unequivocally requires perceiving them as sentient. The relatively universal agreement in the world of science of ethical and legal protection so far prevails to cephalopods and decapods [21].

The concept of evolutionary inclusive ethics [22], postulating the inclusion of invertebrates in the sphere of ethical protection on the basis of recognition of their high degree of sensitivity and the presence of advanced cognitive abilities, is met with a particular response among specialists in animal ethics. The authors argue that the near-total exclusion of invertebrates from ethics-based science policy is not due to the current state of scientific knowledge, but is mainly due to: (i) a naive reading of evolutionary theory that invertebrates are a lower category than vertebrates; (ii) the a priori and false assumption that small brains cannot provide advanced ways of knowing or feeling; (iii) human biases and cognitive-affective biases (e.g., feelings of disgust) that distort moral judgments. The difficulties in initiating practical changes regarding the welfare of invertebrates are summed up by Mather in the introductory article of this volume [17], where she writes that the traditional anthropocentric approach to invertebrates is based on: "our lack of knowledge, our negative attitudes and our misunderstanding of their cognitive abilities..". There is both a logical and scientific basis for including at least some invertebrates in considerations of ethical eligibility. In particular, consistency in the moral treatment of non-human animals is essential: the same characteristics, criteria and reasoning that justify moral protection for vertebrates should serve to extend similar protection to at least some invertebrates. This concept is widely commented on in the world of specialists in animal rights and animal minds [18,20] and it seems to also apply to at least some aquatic invertebrates used in biomonitoring. It seems that an effective way to resolve the ethical conflicts that arise during the implementation of biomonitoring procedures may be the application of the Animal Sentience Precautionary Principle [23]. According to it, doubts about the existence of sentience in a given animal should not lead the researcher to limit the ways of avoiding harm to him. The author proposes to assume the existence of the ability to feel in all representatives of species closely related to the animal, for which it was confirmed experimentally—this also applies to many species of freshwater insects and crustaceans used in biomonitoring.

Some ethicists, however, question the sense of using criteria based on comparing the cognitive abilities of different animals when granting moral rights, treating it as a manifestation of "neo-speciesism" [24]. According to such concepts, it is the welfare of each individual, and not "ingenuity" and relative similarity to human behavior, which should be the basis for moral protection [25].

3. Conclusions

I. Freshwater invertebrates used in the most common biomonitoring methods are still treated as non-sentient beings, thus are devoid of intrinsic value and therefore do not require any form of ethical protection.

II. This leads to misuse, the most common of which is excessive suffering and excessive mortality during sampling. Most often, it takes the forms of non-selective bulk sampling and the use of sub-sampling procedures with the result that the number of animals killed exceeds, sometimes many times, the numbers necessary to properly apply the ecological status assessment protocols.

III. The parallel classification of freshwater animals used in biomonitoring into different categories with different ethical status leads to inevitable ethical conflicts (see Figure 2).

IV. It is difficult to assess the ecological effects of such treatment of free-living populations of freshwater invertebrates, but many individuals of protected species also fall victim to such biomonitoring.

V. The search for improvements and new methods of biomonitoring seems to be necessary and obvious for many specialists. Especially, the development of efficient, precise and reliable methods based on environmental DNA (e-DNA) analysis seems to be not only a breakthrough in biomonitoring, but also an important step towards a significant improvement in the welfare of aquatic invertebrate.

4. Future Directions

The need for change in the approach to the use of invertebrates in biomonitoring seems obvious, both on the part of scientists [26] and practitioners [11]. The discussed problem is a model example of a fundamental conflict between environmental ethics and animal ethics [27]. It seems to be a difficult problem to demonstrate, in accordance with ethics, that the death and suffering of many free-living animals, for which there are logical and scientific premises that they are sentient, can be justified by maintaining the species and ecosystems in a good ecological condition [1]. The benefits of biological monitoring are enormous and undeniable, but raise ethical conflicts when used as a justification for this type of abuse. It seems likely that the lack of generally accepted ethical standards for killing insects may affect personal attitudes related to ethical sensitivity and thus individual behavior [28]. The priorities underlying modern monitoring procedures appear now to be obsolete because they do not take into account the ethical aspect of killing sentient individuals. Large scientific projects in which terrestrial insects are monitored for biodiversity assessment also result in significant mortality [29]—tens of millions specimens). However, they differ from the results discussed in this paper by a very important feature. Namely, unnecessary mortality was reduced there to a minimum—all individuals are by definition identified with maximum accuracy by specialists or archived for future identification. The goal is to know biodiversity, so regardless of the number of individuals killed, the death of each of them matters. In the case of the discussed data on the ecological quality assessment, the majority of individuals are intentionally, but unnecessarily, killed and are later not used for anything. This kind of ethical insensibility is also in contradiction with the more and more commonly formulated proposals to grant ethical value not only to individuals but also to biodiversity, as such [30]. An increasing amount of evidence from experimental research is revealing that many groups of invertebrates have complex behavior and advanced mental potential: substantial perceptual ability, feelings and emotions, pain perception, long- and short-term memory [31], learning abilities, cognitive perception and individual differentiation of behavioral types [32,33]. The view that any activity that causes pain or death in sentient animals should be limited and justified only if the benefits to other organisms or the ecosystem as a whole are significant is becoming more and more accepted [20].

Reducing the negative impact of scientific research, including environmental research, on invertebrates has been postulated for years from ethical positions. Therefore, alternatives to the lethal methods of biomonitoring widely used so far are also sought. Non-lethal methods based on the composition of macroinvertebrates are very few and consist of the morphological identification of digitally recorded images of animals [34–36]. Projects of the photographic identification of various groups of terrestrial and marine invertebrates are being tested, e.g., [37]. For many years, methods have been developed to assess the impact of various anthropogenic stressors on freshwater ecosystems based on the analysis of biomarkers (stress and enzymatic responses, endocrine disruptors, trophic tracers, energy metabolites, genotoxic indicators, histopathological and behavioral alterations and genetic markers) from various groups of invertebrates [38].

The easiest way to avoid most of the described ethical conflicts is a complete resignation from the use of animals in biomonitoring and limiting it to the analysis of non-sentient organisms only: microorganisms (bacteria, algae, protists) and aquatic plants [11]. A reduction in mortality during traditional, routinely used methods of the biomonitoring of freshwater environments can be achieved by the following modifications:

1. Pre-sampling. Before the actual biological sample is taken, a preliminary sample should be taken near the test site, e.g., twenty-five percent of the appropriate sample, in order to estimate the density of invertebrates. Depending on the results of this initial count, the appropriate sample required can be estimated, covering only an area adequate to obtain the final number of animals needed for the assessment. Only those

animals will be killed and preserved. This procedure can be simplified by the use of photographs and videos along with image analysis software (using self-learning systems based on artificial neural networks) that automatically recognizes and counts objects.

2. A similar procedure can be carried out by using modular artificial substrates placed at the sampling point several weeks in advance [32]. Depending on the preliminary assessment, the fauna would be collected from an appropriate number of modules to obtain the final number of animals in the sample close to the minimum required.
3. The application of numerical analysis procedures such as rarefaction, and their use to create empirical saturation curves, will allow the estimation of the maximum abundance of macrobenthos in a sample necessary for a reliable estimate of taxonomic richness [10].

Non-lethal methods used to collect animal body fragments, their tissues or gametes are usually based on procedures that cause stress and suffering in the tested animals [39,40]. They were developed to reduce the mortality, suffering and discomfort of the analyzed individuals in comparison with the methods used so far, therefore they should be considered as activities conducive to animal welfare; however, the ethical assessment of such activities should be approached with great caution. The term 'welfare' used to refer to free-living animals is ambiguous. Especially when used in relation to invertebrates used in environmental studies—it may in fact mean [41] simply a reduction in suffering during exploitation or killing and does not take into account any basic vital interests of animals.

Methods using environmental DNA (e-DNA) are probably the most promising and fastest growing alternative, but it should be emphasized that they were not created for ethical reasons and not all of them can be considered "non-lethal". The main motivation for their development was, in fact, the possibility of a quick and effective assessment of the taxonomic composition of organisms in the environment without the need for time-consuming and labor-intensive morphological identification based on expert knowledge. E-DNA can broadly be defined as the genomic DNA of multiple organisms obtained directly from the environment. This can include approaches that focus on extracting the DNA of taxa presently residing within the sampled matrix, or the collection of degraded fragments of DNA that persist within the environment [42]. Specialists emphasize very rapid development and great prospects for methods based on e-DNA [43], and predict a huge increase in their importance for the purposes of biomonitoring [44]. The effectiveness of e-DNA-based methods in determining the taxonomic composition of organisms in monitoring is still widely discussed. Enthusiastic opinions announcing the rapid replacement of traditional methods [45,46] must face counterarguments pointing to their serious limitations and low reliability [47,48]. Various biological and technical limitations still impede the implementation of the environmental genomics for routine monitoring applications. These limitations mainly stem from the fact that the methods sample fundamentally different units of presence (molecules vs. individuals), resulting in different biases affecting richness, abundance and taxonomic composition [41,49]. The richness of "molecular species" should not be considered analogous to species richness [44,50]. The biggest problem associated with the use of such methods in routine biomonitoring still seems to be the reliable reconstruction of the number or at least the relative abundance of individual taxa in the environment based on DNA fragments found in an environmental sample [51,52]. Certainly, the development of methods based on e-DNA analysis leading to an effective and reliable reading of the quantitative taxonomic composition of aquatic invertebrates with appropriate taxonomic resolution will not only be a breakthrough in biomonitoring, but also an important step towards a significant improvement in the welfare of aquatic invertebrates.

Funding: This research received no external funding.

Institutional Review Board Statement: Not applicable.

Informed Consent Statement: Not applicable.

Conflicts of Interest: The author declares no conflict of interest.

References

1. Rawles, K. Biological Diversity and Conservation Policy. In *Philosophy and Biodiversity*; Oksanen, M., Pietarinen, J., Eds.; Cambridge University Press: Cambridge, UK, 2004.
2. Convention on Biological Diversity. 1992. Available online: https://www.cbd.int/convention/text/ (accessed on 13 May 2016).
3. Water Framework Directive. Directive 2000/60/EC of the European Parliament and of the Council of 23 October 2000; Establishing a Framework for Community Action in the Field of Water Policy. 2000. Available online: https://eur-lex.europa.eu/legal-content/EN/TXT/?uri=CELEX%3A32000L0060 (accessed on 20 November 2014).
4. Bonada, N.; Prat, N.; Resh, V.; Statzner, B. Developments in aquatic insect biomonitoring a comparative analysis of recent approaches. *Ann. Rev. Entomol.* **2006**, *51*, 495–523. [CrossRef] [PubMed]
5. Czachorowski, S.; Buczyński, P. Zagrożenia i ochrona owadów wodnych w Polsce. *Wiad. Entomol.* **1999**, *18* (Suppl. S2), 95–120, (In Polish, English Summary).
6. Sánchez-Bayo, F.; Wyckhuys, K.A. Worldwide decline of the entomofauna: A review of its drivers. *Biol. Cons.* **2019**, *232*, 8–27. [CrossRef]
7. Hallmann, C.A.; Sorg, M.; Jongejans, E.; Siepel, H.; Hofland, N.; Schwan, H.; Stenmans, W.; Müller, A.; Sumser, H.; Hörren, T.; et al. More than 75 percent decline over 27 years in total flying insect biomass in protected areas. *PLoS ONE* **2017**, *12*, e0185809. [CrossRef]
8. Cantonati, M.; Poikane, S.; Pringle, C.M.; Stevens, L.E.; Turak, E.; Heino, J.; Richardson, J.S.; Bolpagni, R.; Borrini, A.; Cid, N. Characteristics, Main Impacts, and Stewardship of Natural and Artificial Freshwater Environments: Consequences for Biodiversity Conservation. *Water* **2020**, *12*, 260. [CrossRef]
9. Suter, G.W.; Cormier, S.M. Why care about aquatic insects: Uses, benefits, and services. *Integr. Environ. Assess. Manag.* **2015**, *11*, 188–194. [CrossRef]
10. Koperski, P. Local variability, human factor or vague procedure? Searching for the reasons of excessive mortality in free living aquatic insects, resulting from biological monitoring. *J. Insect Conserv.* **2023**, *27*, 589–599. Available online: https://link.springer.com/article/10.1007/s10841-023-00482-y (accessed on 2 June 2023). [CrossRef]
11. Koperski, P. Freshwater Invertebrates—Neglected Victims of Biological Monitoring: An Ethical View. *Ethics Environ.* **2022**, *27*, 29–57. [CrossRef]
12. Kolada, A. (Ed.) *Podręcznik do Monitoringu Elementów Biologicznych i Klasyfikacji Stanu Ekologicznego Wód Powierzchniowych. Aktualizacja Metod*; Biblioteka Monitoringu Środowiska: Warsaw, Poland, 2020; (In Polish, English Version). [CrossRef]
13. Friberg, N.; Sandin, L.; Furse, M.T.; Larsen, S.E.; Clarke, R.T.; Haase, P. Comparison of Macroinvertebrate Sampling Methods in Europe. In *The Ecological Status of European Rivers: Evaluation and Intercalibration of Assessment Methods*; Furse, M.T., Hering, D., Brabec, K., Buffagni, A., Sandin, L., Verdonschot, P.F.M., Eds.; Springer: Dordrecht, The Netherlands, 2006.
14. Escribano, N.; Oscoz, J.; Galicia, D.; Cancellario, T.; Durán, C.; Navarro, P.; Ariño, A.H. Freshwater macroinvertebrate samples from a water quality monitoring network in the Iberian Peninsula. *Sci. Data* **2018**, *5*, 180108. [CrossRef]
15. AQEM Consortium. Manual for the Application of the AQEM System. In A Comprehensive Method to Assess European Streams Using Benthic Macroinvertebrates, Developed for the Purpose of the Water Framework Directive; Version 1.0. February. 2002. Available online: https://www.eugris.info/displayproject.asp?Projectid=4422 (accessed on 28 September 2006).
16. Koperski, P. The asymmetry of ethical obligations or the unequal treatment of various groups of animals in scientific research. *Zoophilol. Pol. J. Anim. Stud.* in press. Available online: https://journals.us.edu.pl/index.php/ZOOPHILOLOGICA/authorDashboard/submission/13314 (accessed on 2 June 2023).
17. Mather, J.A. Ethics and invertebrates: The problem is us. *Animals* **2023**, in press.
18. Baracchi, D.; Baciadonna, L. Insect sentience and the rise of a new inclusive ethics. *Anim. Sentience* **2020**, *29*, 18. [CrossRef]
19. Perry, C.J.; Baciadonna, L. Studying emotion in invertebrates: What has been done, what can be measured and what they can provide. *J. Exp. Biol.* **2017**, *220*, 3856–3868. [CrossRef] [PubMed]
20. Broom, D.M. Brain complexity, sentience and welfare. *Anim. Sentience* **2020**, *29*, 27. [CrossRef]
21. Jakopovich, D. The UK's Animal Welfare (Sentience) Bill Excludes the Vast Majority of Animals: Why We Should Expand Our Moral Circle to Include Invertebrates. 2021. Available online: https://onlineacademiccommunity.uvic.ca/asri/2021/10/17/ (accessed on 2 June 2023).
22. Mikhalevich, I.; Powell, R. Minds without spines: Evolutionarily inclusive animal ethics. *Anim. Sentience* **2020**, *29*, 329. [CrossRef]
23. Birch, J. Animal sentience and the precautionary principle. *Anim. Sentience* **2017**, *16*, 17. [CrossRef]
24. Dunayer, J. *Speciesism*; Ryce Publishing: Singapore, 2004.
25. Lejman, J. Etyka zwierząt w świetle idei zrównoważonego rozwoju (Animals' Ethics in the Light of the Idea of Sustainable Development). *Probl. Ekorozw.* **2006**, *1*, 99–105, (In Polish, English Version).
26. Lövei, G.L.; Ferrante, M.; Möller, D.; Möller, G.; Vincze, É. The need for a (non-destructive) method revolution in entomology. *Biol. Conserv.* **2023**, *282*, 110075. [CrossRef]
27. Faria, C.; Paez, E. It's Splitsville: Why animal ethics and environmental ethics are incompatible. *Am. Behav. Sci.* **2019**, *63*, 1047–1060. [CrossRef]
28. Simaika, J.P.; Samways, M.J. Biophilia as a universal ethic for conserving biodiversity. *Conserv. Biol.* **2010**, *24*, 903–906. [CrossRef]

29. Karlsson, D.; Hartop, E.; Forshage, M.; Jaschhof, M.; Ronquist, F. The Swedish Malaise trap project: A 15 year retrospective on a countrywide insect inventory. *Biodiv. Data J.* **2020**, *8*, e47255. [CrossRef]
30. Wilson, E.O. Biophilia and the Conservation Ethic. In *Evolutionary Perspectives on Environmental Problems*; Mysterud, I., Ed.; Routledge: Oxford, UK, 2017; pp. 250–258.
31. Broom, D.M. The Welfare of Invertebrate Animals such as Insects, Spiders, Snails and Worms. In *Animal Suffering: From Science to Law, International Symposium*; van der Kemp, T.A., Lachance, M., Eds.; Éditions Yvon Blais: Paris, France, 2013; pp. 135–152.
32. Perry, C.J.; Barron, A.B.; Cheng, K. Invertebrate Learning and Cognition: Relating Phenomena to Neural Substrate. *WIRE Cognit. Sci.* **2013**, *4*, 561–582. [CrossRef]
33. Carere, C.; Mather, J.A. Why Invertebrate Welfare? In *The Welfare of Invertebrate Animals. Animal Welfare*; Carere, C., Mather, J., Eds.; Springer: Berlin/Heidelberg, Germany, 2019.
34. Karasek, T.; Koperski, P. NoMBSI: A New, Non-lethal Method for Benthos Sampling and Identification for Use in Biological Monitoring of Flowing Waters: Preliminary Results. *Hydrobiologia* **2015**, *751*, 215–227. [CrossRef]
35. Koperski, P. Hydrological instability of ponds reduces functional diversity of freshwater molluscs in protected wetlands. *Wetlands* **2022**, *42*, 42. [CrossRef]
36. Ärje, J.; Melvad, C.; Jeppesen, M.R.; Madsen, S.A.; Raitoharju, J.; Rasmussen, M.S.; Iosifidis, A.; Tirronen, V.; Gabbouj, M.; Meissner, K.; et al. Automatic image-based identification and biomass estimation of invertebrates. *Meth. Ecol. Evol.* **2020**, *11*, 922–931. [CrossRef]
37. Pech, D.; Condal, A.R.; Bourget, E.; Ardisson, P.L. Abundance estimation of rocky shore invertebrates at small spatial scale by high-resolution digital photography and digital image analysis. *J. Exper. Mar. Biol. Ecol.* **2004**, *299*, 185–199. [CrossRef]
38. Colin, N.; Porte, C.; Fernandes, D.; Barata, C.; Padrós, F.; Carrassón, M.; Monroy, M.; Cano-Rocabayera, O.; de Sostoa, A.; Piña, B.; et al. Ecological relevance of biomarkers in monitoring studies of macro-invertebrates and fish in Mediterranean rivers. *Sci. Total Environ.* **2016**, *540*, 307–323. [CrossRef]
39. Leland, J.C.; Furse, J.M. Potential utility of haemolymph analysis in non-lethal conservation studies on threatened Australasian freshwater crayfish: Portability and practicality. *Crust. Res.* **2012**, *7*, 85–93. [CrossRef] [PubMed]
40. Beaver, C.E.; Geda, S.R.; Johnson, N.A. Standardizing a non-lethal method for characterizing the reproductive status and larval development of freshwater mussels (Bivalvia: Unionida). *J. Vis. Exp.* **2019**, *152*, e60244.
41. Wahltinez, S.J.; Stacy, N.I.; Hadfield, C.A.; Harms, C.A.; Lewbart, G.A.; Newton, A.L.; Nunamaker, E.A. Perspective: Opportunities for advancing aquatic invertebrate welfare. *Front. Vet. Sci.* **2022**, *9*, 973376. [CrossRef] [PubMed]
42. Chariton, A.A. Environmental (e) DNA in the aquatic sciences: The CATG is now well and truly out of the bag. *Mar. Freshwat. Res.* **2023**, *74*, i–v.
43. Bohmann, K.; Evans, A.; Gilbert, M.T.P.; Carvalho, G.R.; Creer, S.; Knapp, M.; Yu, D.W.; De Bruyn, M. Environmental DNA for wildlife biology and biodiversity monitoring. *Trends Ecol. Evol.* **2014**, *29*, 358–367. [CrossRef]
44. Pawlowski, J.; Kelly-Quinn, M.; Altermatt, F.; Apothéloz-Perret-Gentil, L.; Beja, P.; Boggero, A.; Borja, A.; Bouchez, A.; Cordier, T.; Domaizon, I.; et al. The future of biotic indices in the ecogenomic era: Integrating (e) DNA metabarcoding in biological assessment of aquatic ecosystems. *Sci. Total Environ.* **2018**, *637*, 1295–1310. [CrossRef] [PubMed]
45. Pawlowski, J.; Bonin, A.; Boyer, F.; Cordier, T.; Taberlet, P. Environmental DNA for biomonitoring. *Mol. Ecol.* **2021**, *30*, 2931–2936. [CrossRef] [PubMed]
46. Ji, F.; Han, D.; Yan, L.; Yan, S.; Zha, J.; Shen, J. Assessment of benthic invertebrate diversity and river ecological status along an urbanized gradient using environmental DNA metabarcoding and a traditional survey method. *Sci. Total Environ.* **2022**, *806*, 150587. [CrossRef]
47. Leese, F.; Altermatt, F.; Bouchez, A.; Ekrem, T.; Hering, D.; Meissner, K.; Mergen, P.; Pawlowski, J.; Piggott, J.J.; Rimet, F.; et al. DNAqua-Net: Developing new genetic tools for bioassessment and monitoring of aquatic ecosystems in Europe. *RIO* **2016**, *2*, e11321. [CrossRef]
48. Cordier, T.; Alonso-Sáez, L.; Apothéloz-Perret-Gentil, L.; Aylagas, E.; Bohan, D.A.; Bouchez, A.; Chariton, A.; Creer, S.; Frühe, L.; Keck, F.; et al. Ecosystems monitoring powered by environmental genomics: A review of current strategies with an implementation roadmap. *Mol. Ecol.* **2021**, *30*, 2937–2958. [CrossRef]
49. Seymour, M.; Durance, I.; Cosby, B.J.; Ransom-Jones, E.; Deiner, K.; Ormerod, S.J.; Colbourne, J.K.; Wilgar, G.; Carvalho, G.R.; de Bruyn, M.; et al. Acidity promotes degradation of multi-species environmental DNA in lotic mesocosms. *Commun. Biol.* **2018**, *1*, 4. [CrossRef] [PubMed]
50. Gleason, J.E.; Elbrecht, V.; Braukmann, T.W.; Hanner, R.H.; Cottenie, K. Assessment of stream macroinvertebrate communities with eDNA is not congruent with tissue-based. *Mol. Ecol.* **2021**, *30*, 3239–3251. [CrossRef]

51. Shackleton, M.E.; Dafforn, K.A.; Murphy, N.P.; Greenfield, P.; Cassidy, M.; Besley, C.H. How does molecular taxonomy for deriving river health indices correlate with traditional morphological taxonomy? *Ecol. Indic.* **2021**, *125*, 107537. [CrossRef]
52. Hering, D.; Borja, A.; Jones, J.I.; Pont, D.; Boets, P.; Bouchez, A.; Bruce, K.; Drakare, S.; Hänfling, B.; Kahlert, M.; et al. Implementation options for DNA-based identification into ecological status assessment under the European Water Framework Directive. *Water Res.* **2018**, *138*, 192–205. [CrossRef]

MDPI AG
Grosspeteranlage 5
4052 Basel
Switzerland
Tel.: +41 61 683 77 34

Animals Editorial Office
E-mail: animals@mdpi.com
www.mdpi.com/journal/animals

www.ingramcontent.com/pod-product-compliance
Lightning Source LLC
Chambersburg PA
CBHW041951260926
51850CB00019B/2417